Collaboration with Potential Users for Discontinuous Innovation

Martin Hewing

Collaboration with Potential Users for Discontinuous Innovation

Experimental Research on User Creativity

Martin Hewing
Universität Potsdam
Potsdam, Germany

Dissertation Universität Potsdam 2013

ISBN 978-3-658-03752-9   ISBN 978-3-658-03753-6 (eBook)
DOI 10.1007/978-3-658-03753-6

The Deutsche Nationalbibliothek lists this publication in the Deutsche Nationalbibliografie; detailed bibliographic data are available in the Internet at http://dnb.d-nb.de.

Library of Congress Control Number: 2013949761

Printed on acid-free paper

Springer Gabler is a brand of Springer DE.
Springer DE is part of Springer Science+Business Media.
www.springer-gabler.de

I dedicate this dissertation to my parents, who supported and encouraged me in all my projects in an outstanding way.

Abstract

Creativity and innovation are important drivers of economic welfare and growth in contemporary societies. Collaborating with and learning from users in the early phase of the innovation process has been considered a successful approach to stimulate those creative sparks for organizations. However, the idea of users as innovators has also invoked critical responses especially in the context of innovations that are discontinuous to dominant designs. Discontinuous innovations result in market and/or technological deviations from the state of the art. Their development is hallmarked by prevalent uncertainties, such as the paradox of building market competences in yet ill-defined market structures. Therefore, some researchers have recently advocated the collaboration with potential users, who are currently not served by an organization. Those users at the peripheries are perceived to contribute more novel information, by which they better reflect shifts in needs and behavior than current users at the center. However, no systematic inquiry into the unique content of their contribution and how it is different to the one of current users has been carried out. Conceptually, the micro processes of construction and utilization of knowledge in collaboration with potential users remain poorly understood. This leads to the recourse of organizations to either traditional market research methods or a sole focus on current users in practice. A better understanding of the unique merits of collaboration with potential users will encourage organizations to extend their market competences to the edges of their own domain. The overall research objective of this dissertation is to validate the merit of collaboration with potential users in creative exploration and conceptually explore how it is embodied in collaborative communication.

To ensure comprehensibility, a mixed method research was conducted, applying a qualitative and quantitative experimental design. The quantitative validation shows that ideas from potential users yield higher newness as compared to the ones of current users, in cases in which they have collaborated with someone experienced. The discontinuity results from a high perceived level of surprise embedded within the idea, which does not stimulate significant technological discontinuity. The qualitative inquiry shows that contributions by potential users are collaterally enriched by their social, cultural and emotional values, emphasizing the search for a meaningful problem in the context of their lives before thinking in terms of problem solutions. Potential users' willingness to improvise and perceive constructive critique from current users as a challenge leads to a series of ad-hoc changes to initial ideas which foster transformation.

The results imply that potential users need structural alignment and access to domain related knowledge in order to perform a creative process which extracts and explicates the essences of their stimuli. Moreover, cross-fertilization arising as a result of collaboration between a potential and a current user is asymmetric due to differences in the nature of the respective knowledge they apply and their personal objectives. The results confirm established theories particular from creativity research in the context of collaboration with users, but also challenge conventional theories and guidelines in organizational exploration. If supportive conditions are met, collaboration with potential users stimulates value co-creation and opportunity recognition. Organizations can learn how to leverage their technological competences in new application scenarios reflecting shifts in behavior of societies that might become more attractive to a larger amount of people over time. Tolerating unpredictability and improvisation are key factors therein.

Table of Contents (overview)

Table of Contents (detailed)

Index of Tables

Index of Figures

1 Overview

1.1 Motivation and Contribution

There is little doubt that creativity and innovation are important drivers of economic welfare and growth in contemporary societies (e.g. Abernathy & Clark, 1985; Glaveanu, 2011; Hennessy & Amabile, 2010). Markets and technologies undergo rapid transformations and often change what rallies around them in radical ways (e.g. Chandy & Tellis, 2000; Christensen, 1997, 2006; Hamel & Prahalad, 1991; Tushman & Anderson, 1986). Though discovery hardly seems to be implementable as a predictable and repeatable process, being creative and innovative has become an imperative for organizations in most domains (Dougherty, 1992). Especially high-technology markets, in which established market boundaries are dissolving, currently face a multitude of disruptive changes. Consequently, organizations not only need to extend their competences to the outside of their own institutional boundaries, but also to the periphery of their own domain and beyond (Faems et al., 2005). Organizations that succeed shifts in markets and technology often base their business on radically different competences to what incumbents before were holding on to (Christensen, 1997; Danneels, 2002; Henderson, 2006; Tushman & Anderson, 1986). The customer competence of organizations is therefore crucial insofar as it enables or restricts organizations in their future innovative activities (Danneels, 2002; Faems et al., 2005). Especially in the early stages of the innovation process a proactive as opposed to reactive market approach has thus proven to be particularly successful (Narver et al., 2004; Witell et al., 2011).

The integration of users into the early innovation process of organizations is widely acknowledged as an effective approach to detect environmental changes and act upon them (e.g. Bharadwaj et al., 2012; Foxall, 1989; Kristensson et al., 2004; Lettl et al., 2006; Salomo et al., 2003; von Hippel, 1986). Many contemporary innovative organizations practice the involvement of users early on (Nijssen et al., 2012; Prahalad & Ramaswamy, 2000, 2004). However, the idea of users as innovators has also in certain instances invoked critical responses especially in the context of innovations that are discontinuous to the dominant design (Callahan & Lasry, 2004; Christensen & Bower, 1996; Ulwick, 2002). Innovations can be differentiated with regard to dimensions reflecting changes, e.g. changes in the need addressed or with regard to the technology used (Veryzer, 1998). Discontinuous innovations are thereby defined as innovations which result in a market and/or technological deviation from the current standard on a macro level (Garcia & Calantone, 2002). The changes involved affect a domain or industry as opposed to a single firm or firm's customers. New organizational competences are required to address these changes. Serving the requests of current users is conceived as limiting the diversity of strategic choices and the path towards incremental improvements (Christensen, 1997; Lynn et al., 1996; van der Panne et al., 2003). The balance between these opposing stances is based on the appreciation that early user involvement fosters the understanding of people's lives and needs, rather than providing a source for fully developed ideas which no longer require creativity on the part of the organization in question (Day, 1999; Lei & Greer, 2003; Leonard & Rayport, 1997; Visser et al., 2007). In the light of discontinuous innovations, some researchers have recently advocated the involvement of potential users, who are currently not served by an organization or who remain in niche markets (Danneels, 2003; Day, 1999; Christensen, 2006; Gilbert, 2003; Govindarajan et al., 2011; Gruner & Homburg, 2000; Lau et al., 2010). Those users at the peripheries are perceived to contribute more novel information, by which they better reflect shifts in needs and behavior than current users (Chandy & Tellis, 1998). Approaching potential users addresses the somewhat paradoxical challenge of building market competences for yet ill-defined market structures. However, no systematic inquiry has yet been carried out into the unique content of their contributions and how to extract it. Conceptually, the processes of construction and utilization of information acquired through collaboration with users in general and with potential users in particular remain poorly understood (Bogers et al., 2010; Greer & Lei, 2012; Kristensson et al., 2008; Ojanen & Hallikas, 2009; Tzeng, 2009). As a result, the Marketing Science Institute (MSI) declared advanced studies of social and cultural user behavior in collaboration to be a top research priority for 2012 (Bharadwaj et al., 2012). Focusing particularly on the collaboration with potential users in the context of exploratory search for discontinuous innovations, this dissertation addresses the following overall questions: *What is the merit of potential user collaboration in organizational exploration regarding content, impact and antecedents, and how is it embodied in the process of collaborative communication?*

Since an insight or idea is the essential building block for successful future innovations (Henard & Szymanski, 2001), the inquiry into the merits and also the drawbacks of collaboration with potential users is important for researchers and managers alike. Empirical studies concerned with potential user involvement have tackled the question on an organizational level and underlined the effects on the innovation performance of business units (Bonner & Walker, 2004; Chandy & Tellis, 1998; Govindarajan et al., 2011; Hang et al., 2010). These studies, however, can be considered to point to three main areas left unaddressed in them, which it is the purpose of the present research to attend to.

Firstly, a validation of the merit of potential user involvement on the organizational level might not be appropriate in the case of explorative activity which is based on collaboration and interaction. The empirical studies offer but a vague guess of the impetus potential users really exert on the final innovation and its performance in a particular domain. Potential users are perceived as informants who contribute insights for innovative ideas (Danneels, 2003; Day, 1999; Mascitelli, 2000; Vargo & Lusch, 2004). According to this, collaboration is located upstream in the innovation process and mostly pursued by a few key individuals in the organization (Reid & Brentani, 2004). These key individuals need to be deeply immersed in the environment and market, since valuable insights from users might be hidden in their subconscious knowledge and experiences (Lei & Greer, 2003; Mascitelli, 2000; von Hippel, 1994). There is a long distance between an initial pivotal insight and the rollout of a final innovation, especially in large incumbent organizations. Although ideas stemming from a collaborative process constitute, at most, intermediate stages of future innovation, they are a more precise analytical subject on the basis of which to evaluate users' contributions. The analysis of contributed ideas provides clues with regard to whether the discontinuity of an idea proposed by a potential user is different in degree to that of a current user.

Secondly, current empirical studies do not address the constituents of the unique outputs of potential users. It is important to know how their contributions challenge the market and the technological competences of organizations in a domain. The focus is on the domain level (macro) and not on the level of single firms (micro). An innovation which is new to an organization but not to a domain is an incremental improvement (Garcia & Calantone, 2002). The impetus of users in incremental innovation is widely acknowledged and quiet well understood. Two general dimensions of innovativeness are acknowledged by most researchers in innovation management: its discontinuity to current market characteristics and its discontinuity to current technologies (Garcia & Calantone, 2002). Contrasting these two dimensions within the evaluation of ideas from potential and current users provides clues regarding the opportunities to acquire new organizational competences or leverage the current ones through user collaboration. This comparative process sheds clearer light on the precise characteristics ("how much of what") the output of collaboration contains or requires for implementation.

Thirdly, and perhaps most importantly, due to the complexity and uncertainty involved in innovative endeavors, innovation development has become increasingly social and collaborative (Dunbar, 1997; Leonard & Sensiper, 1998; Sawyer, 2007; Vargo & Lusch, 2004). While the results of these empirical studies enforce the importance of collaboration with potential users, they do not explain the unique merit that arises and how it can be understood. Furthermore, they do not emblaze the fertile effect collaboration can have on the aspiration of the involved users. Most often, user involvement is not an individual process, but a collective action. Conceptually it is not apparent how potential users influence and are influenced by the innovative activity. A study on the level of micro processes that pays attention to the collaborative process and the inherent communication therefore fills a real gap in user research in innovation. Special focus is directed towards the notion of *cognitive distance*. Cognitive distance refers to the ratio of unique and shared knowledge or to the similarity of the knowledge bases between individuals and is considered to mediate the learning process (e.g. Nooteboom, 2009). The inquiry provides clues about *how* potential users influence the collaborative process and are themselves affected by the collaboration. Furthermore, crucial antecedents and conditions which support potential users to perform a valuable creative process can be identified.

Additionally, though the idea of involving potential users into the innovation process is postulated abundantly throughout the literature on innovation management, no definition can be found and terms are used interchangeably. Researchers refer to these individuals as non-users (e.g. Christensen, 2006; Christensen & Bower, 1996; Danneels, 2003), potential users (e.g. Bonner & Walker, 2004; Fuchs & Schreier, 2011; Shaw, 1985), new users (e.g. Arnold et al., 2011; Danneels, 2002; Henderson, 2006), emerging users (e.g. Day, 1999; Govindarajan et al., 2011; Hoffmann et al., 2010; Slater & Mohr, 2006), prospective users (Anthony, 2009; Danneels, 2003) or future users (Bond & Houston, 2003; Chandy & Tellis, 1998). An analysis of common or distinctive attributions of these terms draws a detailed picture and a common definition of the potential user to start with. Defining and identifying potential users is, however, complicated, since their identity and needs are likely to be at least partly ill-defined and unknown from the present perspective. A micro level perspective and the acknowledgement of the user's subjectivity thus seem all the more crucial.

The practical implications of the aforementioned contributions can challenge the status-quo of how user involvement is carried out in many organizations today. The primary issue of which users to select for meaningful transformation has been neglected and remains poorly understood (Kristensson & Magnusson, 2010). Organizations often fall back to collaboration with their current users only due to missing guidelines on how to integrate potential users and what to expect from them (Christensen, 2006, 2007). In order to successfully manage creativity in user collaboration, it is necessary to understand the contexts under which users, both current and potential, will tend to give rise to ideas that are useful for discontinuous as opposed to continuous innovations (Audia & Goncalo, 2007). Though creative leaps tend to be rather unpredictable in their content (Campbell, 1960; Litchfield, 2008), there seems to be a lot to learn from studying the co-construction processes leading to their emergence.

The theoretical base of this dissertation is interdisciplinary and I therefore propose to draw on two important paradigms stemming from the psychology of creativity. Creative groups and collaborations have most often been explained by theories drawn from cognitive psychology, e.g. in the literature on organizational behavior. In particular the relation between cognition and creativity looms large and can be referred to as the sociocognitive psychology of creativity (Glaveanu, 2011). Its dominant theories on human problem-solving were part of the cognitive revolution which dramatically changed the way creativity was perceived and studied (e.g. Gardner, 1985; Miller, 2003; Newell et al., 1958). The basic assumption of this approach is that all creative outcome emerges from the minds of individuals (Smith et al., 2006). Creativity as a phenomenon is therefore something attributed to the individual and situated in individual cognition. Research on group creativity (e.g. Paulus & Nijstad, 2003) mitigated the strict individual explanatory approach slightly, but individuals still retain their distinctiveness. The origins of something new and useful in group creativity are therefore still considered as being located in individual cognition and their interactions. Social influences in groups, such as social loafing, conformity, production blocking, downward norm setting (Thompson, 2004), topic fixation, social inhibition (Sawyer, 2007), convergence (Larey & Paulus, 1999), absent sharing of divergent ideas or concerns about the evaluation of others (Paulus et al., 2006) are regarded as external conditions with a tendency to result in a predictable negative outcome of the creative process (Glaveanu, 2011). Only recently has a more socially and culturally rooted paradigm of creativity begun to regain attention coming from social psychology. It is frequently referred to as the sociocultural psychology of creativity (Glaveanu, 2011; Sawyer, 2007). Supporters agree that individuals and groups cannot be studied in isolation but only in situated practice and that creativity is situated in the social sphere (e.g. Bruner, 1990; Cole, 1996; Glaveanu, 2011; Shweder, 1990). Creativity arises from communication which exerts influence on further actions and causes unexpected properties to emerge. Emergent properties are unintended effects of action (Sawyer, 2005). The main difference between the sociocognitive and the sociocultural paradigm is therefore rooted in the basic understanding of the origin of creativity – creativity located in the individual cognition as opposed to located in collective or distributed cognition (Hutchins, 1995). The sociocognitive view finds echo in literature on creativity in innovation, while the sociocultural perspective still receives scant attention (Sonnenburg, 2004). Given the fact that most innovative activity is nowadays carried out in collaboration (Leonard & Sensiper, 1998; Reid & Bentrani, 2004; Shaw, 1985; Sonnenburg, 2004), a sole focus on individualistic approaches will make its essence hard to grasp. Within this dissertation, individual contributions are studied, just as the collaborative process and its emergences. It constitutes, therefore, a first attempt to give reasonable attention to the sociocultural psychology of creativity in user involvement, as well as to strengthen the research domain by a comprehensive approach. Furthermore, whilst there is a backlog of experimental research on creative problem-solving which analyzes the effects of the manipulation of explicit knowledge, studies on the manipulation of tacit knowledge remain scarce (Runco & Sakamoto, 1999). The contributions of users are acknowledged to have a high degree of tacitness (Mascitelli 2000; von Hippel, 1994). Transferring theories from other disciplines without paying attention to the nature of knowledge can lead to false interferences. We therefore examine the nature of knowledge and information that users contribute and how it affects the collaborative process. It is a first approach to strengthen collaborative experimental research within the context of user research.

In sum, despite the managerial importance of a well-structured exploration towards new market opportunities, little attention has been directed towards collaboration efforts within more distant markets on a micro perspective and towards more collective theories which are appropriate to the participative nature of innovative activities. This dissertation thus makes the following three important contributions to the literature on user involvement in early innovation activities:

Conceptual contribution. An integrated definition of the potential user and approaches to classify users as potential users are derived from conceptual and empirical literature on innovation management. Expected merits to arise from potential user collaboration are aligned to theories from creativity research in order to gain theoretical conceptualization. Literature on defining degrees of innovativeness is scanned to align those merits to specific and acknowledged innovation typologies. Validated constructs are reviewed to propose a classification scheme of users which can be operationalized in empirical testing. Finally, significant research gaps in collaboration with potential users are identified. This is conceptual work and will be the topic of article I of this dissertation. It is the basis for an empirical validation and inquiry.

Validation contribution. The outcome of collaboration with potential and current users is compared and contrasted. The focus lies on a comparison of the newness, the surprise and the technological requirements of ideas from potential and current users who engaged in collaborative problem-solving in varying dyadic compositions. This is the objective of article II.

Inquiry contribution. A micro process perspective on communication in collaboration will provide clues regarding the unique contributions of potential user in terms of their problem-solving strategy and content compared to the ones of current users, how they are embodied in the process and what emerges from their collective action. The goal is to elaborate new concepts unique to the research context of collaborative problem-solving with potential users, on the one hand, and to scrutinize propositions from the literature in an iterative manner, on the other. Guidelines of how to collaborate with potential users in organizational practice are derived. These will be of relevance to key individuals in organizations, who are concerned with user involvement. The integration of the three overall contributions leads to a comprehensive picture and moves towards a process and content theory of collaboration with potential users. This is the central goal of article III.

The respective contributions are formalized in the following research questions:

1. Are the ideas contributed by potential users more discontinuous to current market characteristics than those of current users?
2. Do the ideas brought in by potential users constitute a greater challenge to technological competences within a domain than those of current users?
3. Does the cognitive distance between collaborating users influence their ideas and are there differences between potential and current users?
4. How do the contributions and strategies of potential users differ from the ones of current users in creative and collaborative communication?
5. How are both potential and current users influenced by what emerges from the collaboration and with which outcome?
6. What are the supportive antecedents and conditions making the collaboration with potential users salient?

These research questions call for a unique and integrative research design which is described in the following paragraph. For clarification I want to mention that this dissertation analyses processes of direct interaction and collaboration between users, which is often coined as *co-creation* (e.g. Kristensson et al., 2008; Prahalad & Ramaswamy, 2000, 2004). User research is in itself diversified and also encompasses forms such as the user as entrepreneur (e.g. Baldwin et al., 2006) or mass collaborations via marketplaces and communities (e.g. Piller & Walcher, 2006). These latter research streams will not be the focus of this dissertation. Please note that article I and II were developed together with the co-author Katharina Hölzle and therefore the subject *we* is used to refer to the authors of the study.

1.2 Method and Data

When referring to research on collaboration and creativity, Steyaert and Bouwen (2004, p. 143) appropriately mention: "[…] there is no single form of using a specific method, since it is a creative and context-bounded application". The research objective of this dissertation does not allow for standard methods to be used unaltered. For this reason I will elaborate in further detail the methods that have been applied, the data that has been gathered and the process of analysis that has been performed. This part will be similar to the methodological parts of each article, but to explain it here at the start will increase the comprehensibility of the overall conclusion drawn in chapter 1.4 of this overview.

1.2.1 Overall Research Design

Since the research questions require a qualitative and a quantitative design, a multi-method approach was applied which was designed under the principles of mixed methods research (Johnson et al., 2007; Teddlie & Tashakkori, 2011) and triangulation (Denzin, 1989; Miles & Hubermann, 1994; Pratt, 2008). The reconstructive linkage of the results from different methods delivers a rich and comprehensive picture from different perspectives. Miles & Huberman (1994) distinguish four models of methodological triangulation. One of the models considers a quantitative method as being complementary to extensive qualitative research. This study is considered to follow this design. Johnson et al. (2007) first coined the term *qualitative mixed* for studies in which the qualitative inquiry dominates. The main research goal is to explore and uncover differences in the contributions and approaches of potential users within explorative tasks compared to current users and emergences arising from the process of collaboration. A qualitative setting decoding the communication process between people and how individual utterances are synthesized as the collaboration moves forward is required (Samra-Fredericks, 2004; Sawyer, 2005). In-depth analysis of the process is then complemented by a quantitative evaluation and comparison of the output (e.g. ideas) of collaborative processes. The research objective is comprehensive insofar as the individual, the collective process and the output are all connected grains of focus throughout the articles.

In the process of the research, we conducted idea discussions, in which potential and current users were asked to collaborate and to come up with new ideas within a given domain. In total 68 participants were invited to take part. The discussions were conducted in dyads in order to diminish social influences of groups and to be able to deconstruct the collective action precisely (Rubenson & Runco, 1995; Sawyer, 2005). Three dyadic compositions were possible: (1) a potential user discussing with a potential user (inexperienced homogeneous dyad), (2) a potential user discussing with a current user (heterogeneous dyad) and (3) a current user discussing with a current user (experienced homogeneous dyad). The domain of the task was music applications on mobile devices (e.g. mobile phones, tablets). There are many non-users within this domain. A large proportion, however, will have some affiliation with music as such, and can thus be considered potential users according to the definition established in the conceptual article I (see chapter 2).

Structured questionnaire. Through the development of a structured screening questionnaire, the participation of both potential and current users was assured. The sampling followed a purposive approach (Given, 2008; Miles & Huberman, 1994). The questionnaire contained domain-dependent and domain-independent constructs and was sent to university networks and to a user panel of the Telekom Innovation Laboratories. The incentive for participation was ten Euro. The domain-dependent constructs were used to classify respondents as a current user, a potential user or as not suitable for the idea discussions. The construct *use experience* (Alba & Hutchinson, 1987) was used as the main classificatory variable. Respondents were asked to indicate whether they used music applications and applications in general, how many music applications they used and how often they used them. Current users were those respondents who indicated that they used music applications on a weekly basis. Respondents who did not use any applications on mobile devices, but indicated a positive tenor towards music applications in general were classified as potential users. Their attitude was identified through the *involvement* construct (Zaichkowsky, 1985) which reflects personal interest, values and needs towards a product or domain. Of each user class, i.e. potential and current users, 34 participants were randomly selected, invited and randomly assigned to dyads. Additionally, the screening questionnaire contained personal inventory scales such as *creativity* (Im et al., 2003; Kirton, 1976), *empathy* (Zeidner et al., 2004) and *problem-solving expertise* and also demographic inquiries about age, gender and occupation. They served as control variables in the quantitative analysis. The idea discussions were conducted within one week and went through six successive steps:

1. ***Start-up.*** The first step involved an introduction of the participants and the researcher. The sessions were tape-recorded from this step on.

2. ***Warm-up.*** The participants were then asked to perform two so-called ice-breaker tasks. The goal of the first task was to reduce the barrier between the participants and the researcher and to become more familiar with each other. Informal communication fosters cohesiveness and is conducive for the output of creative endeavors (Edmondson, 1999). The second task was a short collaborative problem-solving exercise. Participants would thereby lose any fear of talking out loud and become comfortable to clearly articulate their thoughts.

3. ***Idea discussion.*** The idea discussion was the main creative problem-solving task. Participants were asked to collaboratively think of new music applications on mobile devices. The task was first read out to the participants and a print out of the task was available for them to look at again while discussing. The participants were told to interpret *new* as being *new to them*. The task was open-ended and allowed participants to engage in divergent thinking (Guilford, 1968; Runco & Sakamoto, 1999). The participants should clearly verbalize their thoughts to their discussion partners for two reasons. First, sharing ideas is an important component of

collaborative and creative problem-solving (Paulus & Yang, 2000). Second, clear articulation of thoughts enriches the data and uncovers hunches to the germs of contributions and ideas in communication. The idea discussions were terminated after 20 minutes. During this time the researcher was withdrawn from the space of the two participants and taking notes from observation. As a result, participants could not seek a response (e.g. reassurance) from the researcher and were thus free to steer the creative process themselves.

4. *Idea template.* After the collaboration the participants were separated and asked to note one idea on an idea template in written form. This included giving the idea a name and providing a description of its purpose. The individual specification of an idea gave each participant deliberate time to reflect on the ideas after the collaboration. Individual reflection is important for collaboration to be beneficial in terms of output (Paulus & Yang, 2000). In total 68 ideas were contributed.

5. *Unstructured questionnaire.* After completion of the idea template, each participant was asked to fill in an unstructured questionnaire indicating how he or she came to grips with the idea(s) in the collaboration and how he or she felt about the idea discussion. Statements by participants can reveal their subjective feeling of whether the collaboration had led to a change of his or her knowledge and/or perspective.

6. *Debriefing.* The experiment concluded with a short debriefing of the participants as to what the research was about.

Subsequently, each idea was evaluated by an expert panel. The evaluation criteria were based on the Consensual Assessment Technique (Amabile, 1996), with a slight adaptation in order to properly map the discontinuity of ideas in terms of market and technological discontinuity (Garcia & Calantone, 2002). The adaptation was based on established criteria. In detail three criteria were selected for the evaluation, namely *newness* (Amabile, 1996), *surprise* (Jackson & Messick, 1965) and *requirement for new technologies* (Goldenberg et al., 2001). We added surprise, since newness alone can also reflect original but somewhat continuous ideas (e.g. Audia & Goncalo, 2007; Sternberg et al., 2003). Surprise is the expression of unusualness in a product or service judged against dominant market designs. It reflects market discontinuity in a deviation from common expectations (Amabile, 1996; Jackson & Messick, 1965). Furthermore, we used requirement for new technologies (Goldenberg et al. 2001; Veryzer, 1998) instead of the often-used *producibility* criterion. It describes whether organizations need to acquire new technological competences to implement ideas emerging from the edges of the domain or whether existing technological competences can be expanded over established domain boundaries.

The expert panel consisted of six professionals from different sectors of the music and music software industry. Two of the experts had a technical engineering background, another two came from the realm of marketing, and yet another two had a business development background. According to the *consensual assessment technique* (Amabile, 1996), the ideas were to be evaluated relatively to each other. The experts rated the ideas independently of each other and each expert was asked to rate the ideas on all three criteria on a seven-point Likert scale from one (very low) to seven (very high). The intraclass-correlation-coefficient (ICC) of each criterion showed a high consensus of evaluations among the six expert judges, with values equal or above .7 (McGraw & Wong, 1996; Shrout & Fleiss, 1979).

1.2.2 Data Analysis

The research design described above allows one to extract both qualitative and quantitative data. The (1) transcripts of the idea discussions, (2) field notes and (3) statements from the unstructured ex-post questionnaire is data related to the process and collective action and will be analyzed by means of a qualitative approach. The (4) screening questionnaire with personal inventory constructs, (5) the classification of the participants, (6) the dyadic composition they participated in and (7) the expert evaluation of their ideas, taken together, constitute data that relates manipulation and its effect on the output and is analyzed using a quantitative approach.

The method of collection and analysis is thus simultaneously both a qualitative and a quantitative experiment (Clark et al., 2008). From a traditional perspective this might appear slightly paradoxical. The experiment as a research method emanated from natural science, in which the ability to manipulate distinct variables, to measure outcomes objectively and to exert strict control over influencing factors are basic requirements (Bailey, 2008; Cook & Shadish, 1994; Kleining, 1986; Liebert & Liebert, 1995). The researcher manipulates and consequently takes an active part (laboratory experiment). These requirements seem to conflict with those of a qualitative experiment as it is frequently applied in social research and expected to be everyday life-like, open, flexible and directed towards unexpected discovery (Denzin & Lincoln, 2011). The researcher observes and therefore absorbs (field experiment). Kleining (1986) concluded, however, that both activities - to act and to absorb - belong together, are mutually constituent and a strict separation should not be maintained. Furthermore, mixing quantitative justification and qualitative inquiry in experimental group settings is neither new nor unfounded. Teddlie et al. (2008) coined the term *group-case mixed method research*, which integrates ethnographic techniques and experimental designs to strengthen the conclusions through corroboration, explanation, and elaboration. It is also referred to as *experimental ethnography* in which random assignments to experimental and control groups are used in field experiments (Sherman & Strang, 2004; Tweney, 2004). In the case of active collaboration with users in an organizational setting both absorbing and acting coexist. The key individual of the organization on the one hand gives clear instructions to the users and thereby acts, while he or she observes and extracts insights from the emerging interaction. It is both a field setting (everyday-like) and a laboratory setting at the same time. The conduction of small group creative sessions with users in innovative activities is common organizational practice (Prahalad & Ramaswamy, 2004). The participants in these sessions, once instructed by the key individuals of the organization, are given freedom to express their thoughts, can decide in a self-determined manner what to contribute and which dynamic the process can take with regard to its rhythm, the alternation between those involved and its depth. Observing the interaction is like observation in a natural realm of organizational practice. It is, however, a laboratory-like setting as well, since the participants, after being classified as potential or current users, are randomly assigned to collaborative partners. The classification as potential or current users, and the dyadic composition they participate in, constitute the experimental treatments carried out by the researcher. The participants were given clear guidance concerning the task and the goal, the permitted means and the basic rules of social interaction. Furthermore, personality inventory measures serve as control variables, which assure the internal validity of the design. This setting is located in the middle of the continuum between a created and natural collaborative setting (Siever, 1968; Steyaerd & Bouwen, 2004).

Quantitative experiments are a basic method in research in cognitive psychology and relevant to the sociocognitve psychology of creativity (Paulus et al., 2006). They are mostly conducted in order to test hypotheses derived from theory and causal inferences are attributed to the manipulation carried out by the researcher. Qualitative experiments are more often observed in social psychology and in the sociocultural psychology of creativity (Moran & John-Steiner, 2004; Sawyer, 2007; Seddon, 2004). They explore the structure of a (social) phenomenon. In this dissertation, these approaches are combined, which is possible due to the fact that the sociocultural paradigm does not neglect the individual as one of the analytical levels. This is called analytical dualism (Archer, 1995) and renders the connection and mutual benefit of both approaches possible.

The quantitative experiment. A 2x2 between-subject factorial block design was laid out to test systematic differences in the discontinuity of ideas brought about by the experimental treatments. The distinction between potential and current users was labeled *user typology*, referring to differences in their past experience endowment. This is the first experimental factor and independent variable. The other is the *cognitive distance* between the participants of one dyad. Cognitive distance was high in cases where the participant had been in a dyad with an unequal experienced user, e.g. a potential user in discussion with a current user. It is low in those dyadic compositions that were composed of equally (in-)experienced persons, e.g. dyads of potential users. The dependent variable is the *discontinuity* of their individually contributed idea as evaluated by the expert panel, separated into *newness*, *surprise* and *requirement for new technologies*. Since the ideas were contributed individually after the participants engaged in collaboration, four experimental units arose: an idea that was contributed by (1) a potential user who had discussed with a potential user before, (2) a potential user who had discussed with a current user before, (3) a current user who had discussed with a potential user before and (4) a current user who had discussed with a current user before. The differences between the experimental units were tested by means of a factorial analysis of variance (ANOVA). The personal inventory measures and the demographic indicators acquired through the structured questionnaire, served as control variables mitigating the external influences of differences in age, gender, creativity (Kirton, 1976; Im et al., 2003), empathy (Zeidner et al., 2004) and problem-solving expertise indicated by the individuals.

The qualitative experiment. The observation of interaction and emergences in small groups is acknowledged as ethnography of communication (Hymes, 1974; Saville-Troike, 2003) and has been applied in seminal work on collaborative and creative problem-solving (e.g. Dunbar, 1997; Fernández-Cárdenas, 2008). It allows for a comprehensive view of creativity in context and unravels the effects of individual perspectives and their natural evolution in a collaborative effort (Glaveanu, 2011). The transcripts of the dyadic idea discussions and the field notes constituted the core data for the explorative analysis following the principles of *grounded theory* by Glaser & Strauss (1967). Discussions are not conducted to obtain opinions on pre-assembled problems, like in interview methods, but to probe into implicit information and behavior of participants and collectives and generate insights from action (Steyaerd & Bouwen, 2004). The transcripts and interpretations of the researchers were complemented by the statements the participants made within the unstructured questionnaire. Participants could express their thoughts on the collaboration and how they felt during the communication in their own words. The comprehensive analysis of the data contained grains of focus on the individual and the collaboration. Tentative propositions about the merits of collaboration with potential users were established and subsequently integrated into a framework. The framework applies the *paradigm model* by Strauss and Corbin (1996) and depicts a basic social process. Throughout the open and axial coding, constant comparison to existing literature was conducted and incoming data and insights were tested against already coded data. To understand the micro genetic process and identify the germs of the creative sparks, special focus was directed towards unique idea episodes[1] within an idea discussion. To ensure reflexivity and credibility of the category coding, a variety of quality checks were applied, e.g. comparison of coding with other researcher, member checks, frequent debate of codes and interpretations with fellow researchers and the integration of own experience from applied user involvement in innovation. Each derived concept and category will be presented with power and proof quotes (Pratt, 2009).

To sum it up, there are two dimensions that structure the overall analysis of the dissertation which is depicted in Figure 1.1. The first dimension marks the instances which determine the grain of focus of the analysis, with a focus on either the individual or the collaboration and their mutual relation. Then there is the dimension of data, focusing either on the process and communication in the idea discussions or on the impact of its output represented by the evaluation of the idea templates.

Figure 1.1 also depicts the content of each empirical article. Article II focuses on the output of the creative process, analyzing the quantitative data. Article III builds on the qualitative part of the experiment. It follows a dialectical stance (Greene, 2007; Teddlie & Tashakkori, 2011), since within the discussion the findings are related to the results from the quantitative analysis of article II. Article I is conceptual and therefore not reflected in the methodological overview of the empirical research in Figure 1.1. The article in chapter 5 is a continuation of prior research which was done and published during the course of the dissertation. It is not directly related to the articles before and therefore only included as an appendix to this manuscript.

---

[1] We refer to parts of the discussion in which the focus was on the elaboration of an idea or stimulus as idea episode.

Figure 1.1: Overview of Mixed Method Approach

	Process	Output
Individual	**Grounded Theory** Data • Idea Discussion Transcripts • Unstructured Questionnaire • Field Notes Goal • Contrasting Contributions • Explore Patterns	**Factorial ANOVA** Data • Expert Evaluation of Ideas • Experimental Treatments • Structured Questionnaire Goal • Impact of Experimental Effects on Discontinuity of Ideas
	Research Question 4 **Article III (Chapter 4)**	**Research Question 1, 2 & 3** **Article II (Chapter 3)**
Collaboration	**Grounded Theory** Data • Idea Discussion Transcripts • Unstructured Questionnaire • Field Notes Goal • Explore Structure in Emergences • Identify supportive Antecedents	
	Research Question 5 & 6 **Article III (Chapter 4)**	

1.3 Summary of the Articles

I will now summarize the major findings and implications of each single article, before chapter 1.4 draws an overall conclusion from this dissertation.

1.3.1 Article I: Co-Creation with Users at the Edges of Markets

The objective of the first article is to create a theoretical conceptualization of collaboration with potential users, which includes (1) the development of a comprehensive definition of the potential user from innovation management literature, (2) the identification of merits of potential user collaboration expected to emerge within the literature and the correspondence thereof to theories drawn from creativity research and literature on degrees of innovativeness, (3) the identification of schemes to classify users into potential or current users used in the few existing studies, (4) the suggestion of a classification scheme using validated constructs from consumer research and (5) an outline of significant research gaps which call for further research.

A systematic literature review in the field of innovation management uncovered a diversity of terms used to paraphrase a user who is not a current user, but of interest to an organization, such as non-users (e.g. Christensen, 2006; Christensen & Bower, 1996; Danneels, 2003), potential users (e.g. Bonner & Walker, 2004; Fuchs & Schreier, 2011; Shaw, 1985), new users (e.g. Arnold et al., 2011; Danneels, 2002; Henderson, 2006), emerging users (e.g. Day, 1999; Govindarajan et al., 2011; Hoffmann et al., 2010; Slater & Mohr, 2006), prospective users (Anthony, 2009; Danneels, 2003) or future users (Bond & Houston, 2003; Chandy & Tellis, 1998). Most of these terms are used interchangeably and definitions are few. Commonalities between these terms found throughout the literature relate to i) the present non-usage of a product or service by the person in question, which might in some cases be the result of constraints, and ii) the person's positive sentiment and interest towards the particular domain. Similarly to Hauser et al. (2006), the commonalities imply that these people have a genuine interest in the domain and that occasional users with rather low verve or the vast population of non-users are not part of the definition. Potential users are not the current focal point of established organizations, but have the potential to gain importance in the near future (Govindarajan et al., 2011). The advantage of working with potential users is rarely defined in greater detail and mostly expressed indirectly in the converse argument that focusing on current users triggers more of the same (e.g. Danneels, 2003). The implications of this and other statements are discussed and aligned to theories mostly stemming from the psychology of creativity. The merit expected to emerge corresponds well to the definition of discontinuous innovation, which stimulate market and/or technological discontinuities on the macro level of a domain (Garcia & Calantone, 2002). It is necessary to be explicit about the degrees of innovativeness in research, as it prevents confusion of terms used throughout the literature and makes empirical studies and their outcome comparable. Throughout the literature review, it became increasingly obvious that defining who will be important users in the future is as difficult as depicting which technologies will change the competitive landscape. Potential users and their needs will always be partly ill-defined and not well-known from the perspective of the present. Two dimensions based on three validated constructs from consumer research are proposed in order to operationalize the classification of users into potential users in alignment with the former given definition. The construct use experience (Alba & Hutchinson, 1987) is most important as it constitutes knowledge through direct interaction between the user and a product and thereby affects the user's mental models. User knowledge (Park et al., 1994) in contrast is a construct that classifies the individual on a more theoretical expertise level of knowledge (Brucks, 1985). Potential users for instance might lack pristine experience with products or services in a domain, but might have user knowledge through interaction with peers or information seeking behavior. To rule out the vast population of non-users, the involvement construct (Zaichkowsky, 1985) is an appropriate measurement to select those people, who - even though they are non-users - feel affectionate towards the domain. Along these three constructs and their items, an individual can be classified as a potential or current user with different ratios of experience in and needs related to a domain.

In conclusion, the article identifies three significant research gaps from the literature review, which are now addressed within this dissertation. Firstly, the idea of absorbing more discontinuous stimuli for innovation by collaboration with potential users as opposed to current users needs conclusive corroboration at the point of its origin. Secondly, in order to identify leveraging potential, it is important to know how their (salient) contributions challenge both the market and the technological competences of an organization. Thirdly and most intriguingly is the lack of empirical exploration into their salient contributions regarding content and the micro processes of its emergence in communication. An empirical inquiry into the embodiment of the unique contributions of potential users in communication is needed in order to advise organizational practice in exploration. The conceptualization is the basis for a subsequent empirical validation of and inquiry into the merits of potential user collaboration.

1.3.2   Article II: Innovative Ideas through Collaboration with Potential Users

The second article mainly addresses research questions one to three. The focus of the analysis is on the individually contributed outputs - the ideas - and their evaluation after the users have discussed ideas in collaborative dyads. It explores the discontinuity of ideas as being influenced by the *user typology* (potential vs. current user) and the *cognitive distance* (low vs. high) between the discussion partners in the dyads.

The analysis yielded four interesting results defining the merits of user integration more precisely when increased discontinuity of the outcome is desired. First, potential users do contribute ideas that have a significantly higher market newness in cases in which they have collaborated with a current user before. Second, a potential user's ideas are most obviously characterized by an increased level of surprise and unusualness which seems to arise from new properties embedded in the ideas or new contextual relations and does not require technological leaps. There was no significant difference in the requirement for new technologies to pursue these ideas. Third, there were no significant differences between the ideas of potential users from homogeneous dyads and the current user's ideas. Mutual lack of knowledge of the status-quo within the domain and anticipation of common usage scenarios might explain the performance of the homogeneous potential user dyads. Last and perhaps most enlightening is how the analysis showed a strong and significant interaction effect between the independent variables user typology and cognitive distance. This indicates that potential and current users were affected differently by cognitive distance. The effectiveness of mutual cross-fertilization between potential and current users is asymmetric, suggesting that potential users benefit more from the experience of current users than current users from the flexibility of potential users. In the context of user collaboration, two conditions cause this asymmetry. On the one hand, a current user tends to have difficulty understanding and formalizing accurately the tacit knowledge embedded in the contributions of a potential user, while the domain-related knowledge of a current user is more easily understood by potential users. This indicates that in user collaboration, cognitive distance between collaborators with respect to a particular domain leads to unequal learning conditions and required skills.

On the other hand, the personal objectives of a current user will guide his or her selection of the contributed idea. Current users are generally unwilling to dispense with performance characteristics of current products which are highly valued, since after all they address an established need. Their personal objective thus collides with the task objective. Generally, the article shows that the potential user is an attractive collaboration partner in explorative activities, especially when new market opportunities are the desired goal of the organization. Their contribution is original and relates to new usage scenarios and contexts which provide a new benefit to the users without the need for technological leaps. It is not technology but technology-in-use that matters (Paap & Katz, 2004). Discontinuous innovations are not necessarily radical on a technological level (Garcia & Calantone, 2002). They emerge, rather, in the interplay between new and old needs, new and old technologies (Paap & Katz, 2004). Following the competence-based typology of product innovations of Danneels (2002), the collaboration with potential users can leverage technology. The existing technological competence of an organization can be applied to new contexts or different needs. This refers to the delinking of technological competence from established products and contexts, and the relinking thereof to new ones. The leveraging types of innovations are regarded as discontinuous (Garcia & Calantone, 2002), albeit with a medium degree of overall innovativeness, insofar as either technological or market competences largely remain unaltered (Danneels, 2002). In the context of collaboration with potential users, it is technological competence that is challenged to a lesser extent than market or customer competence. Kock et al. (2011) have shown that innovations with this particular trait are commercially most successful.

Furthermore, the article depicts limitations to user collaboration in general and the merits of potential user collaboration in particular. The potential user's salient contributions require structural alignment to the status-quo within the domain that was accessed through the knowledge acquired in heterogeneous dyads. This finding supports the *foundation view* in the interplay between creativity and prior knowledge (Weisberg, 1999), which postulates that prior domain-related knowledge is a basic requirement to be creative within the domain. The foundation view still struggles with the as yet unresolved issue of whether expertise (deep immersion into a domain) or rather familiarity (loose knowledge) leads to better results (Weisberg, 1999). The results presented in this article seem to suggest the latter within the context of user involvement, in which the potential user can absorb basic domain-related information quickly. The finding relates therefore to more contextual approaches of real-time and short-term learning within research on organizational learning (Miner et al., 2001). Furthermore, these findings reveal limits to the cross-fertilization process since a current user does not benefit from or make use of benefits from the collaboration in the same way as a potential user does. These results thus present a more precise picture of the creative impact of the merits that can be expected from collaboration with potential users.

1.3.3   Article III: The Playful Ingenuity of Potential Users in Collaboration - Enriched Compensation and Improvisation

Article III explores how potential users influence collaborative problem-solving as opposed to current users and how both potential and current users are influenced by emergent properties of the collaboration. The findings are transformed into tentative propositions and related to the individual outputs as analyzed in article II.

One main concept emerged throughout the analysis, which was subsumed under the core category *enriched compensation and improvisation*. It is embodied in seven strategies and emergences coded *simplification, critical incident, personal transfer, personal agenda, sense-making, assimilation* and *deliberate debate* and delineates a basic social process, which has roughly two stages.

Enriched compensation. Particularly in the early phase of the idea discussion, the potential user pervades and enriches the communication with his or her emotional, social and cultural values and views. This information originates rather unintentionally from the strategies of the potential user to compensate the domain-related experience gap. The enrichment relates to individually experienced situations, daily life habits and emotions as opposed to artifacts from the domain. One example taken from a total of six different enrichment concepts is the concept labeled *critical incident*. Recently experienced events were an important source of information in order to process the creative task. Such incidents were recalled and shared by the participants. Current users tended to draw on events they experienced with applications, such as flaws or common usage routines. This makes the critical incident object-related. The creative task triggered vivid pictures of recent interactions between the person involved and the application. Although the situation is admittedly recalled, the ideas emerging from it were requirements to existing entities. Potential users carried out a similar mental leap, which was rooted in emotions. The emotions caused the vivid recall of the related situation. This makes their incident context/emotion-related and not directly connected to a particular artifact of a domain. This is how one potential user remembered a recent event:

> Last week I was sitting in the bus and saw someone nodding his head heavily to the music on his headphones. I got excited and curious and would have loved to have been able to listen to his music. [...] an application that could connect me to his player would be a good and communicative idea.

The root of the idea lies in an emotion which emerged from a particular situation. When summarizing the exemplary finding from above more theoretically, it becomes apparent that potential users tend to spend time on finding a problem to solve that is meaningful in the context of their lives, while current users tend to start thinking in terms of solutions more quickly. This leads to the following exemplary tentative proposition:

Proposition Example. While current users' source of stimuli for the creative process tends to be object-related, potential users rely on context-related stimuli and emotions to derive needs and wishes. The former behavior is more likely to lead the collaboration to more obvious problems, while the latter has the potential to formulate a new problem that is rooted in new contexts.

Enriched improvisation. While enriched compensation is behavior which can be traced back to individual contributions, the enriched improvisation part of the core category refers to a process that emerged predominately between potential and current users. Especially in cognitive distant dyads the unintended enrichment of the information was prevalent, as the potential user felt the need to explain himself or herself more thoroughly. He or she spent more time concretizing a problem he or she was trying to address within a context to give meaning or justification to the idea (*sense-making*). In this way, a form of collective cognition can be seen to be established through mutual learning and unlearning (*assimilation*). It directs the collaborative attention to these intangible idea stimuli. One could in fact argue that the assimilation process is a form of group convergence studied in sociocognitive psychology (e.g. Larey & Paulus, 1999).

Indeed, members of groups tend to direct attention to information which its members have in common (e.g. emotional values to which both participants can conform). Here, it is even more so a process of learning and unlearning, which is a property of collaboration, not simply the result of individual participants (Hutchins, 1995; Sawyer, 2005). This process creates a mutually perceived safety zone in which mutual guidance and critique are practiced at the same time and lead to improvisation - the *deliberate debate*. The personal objective of the current user to align unstructured ideas to real-life requirements is perceived as a challenge, not a shattering critique by the potential user. Constructive critique or challenges thus trigger new divergent processes, in which ideas are radically altered to incoming requirements. The participants take comfort in the ill-definition of the task and start to improvise. New information for both members emerges and influences their subsequent actions. Each utterance can mark a small additional idea or a challenge to the previous utterance. The collective cognition evolves. Novelty arises from a transformation of the critique into an unanticipated opportunity (Miner et al., 2001). Improvisation is embodied in (1) cancelling out the critique by altering the idea slightly or (2) redefining the critique and infusing novel meanings. When a current user complained about the limits of mobiles phones in music production due to their small screen sizes, the potential user rapidly infused new meaning to the limitations:

> Sometimes it is the limitations of parameters and functions that spark your musical creativity. A reduction to important functions on a smart phone or tablet could achieve that.

The process and salient output of potential users arise, in other words, from a collaborative effort. It is improvisational, since it is a collateral outcome and not planned experimentation (Miner et al., 2001). Collaboration with potential users is thus not about the statistical probability of getting one good idea out of many, as often practiced in organizational exploration (Goldenberg et al., 1999; Simonton, 1977). It is, rather, about empathy for the unique qualities of the situations and life habits that are best extracted through thorough debate between the members. Performance is therefore to be acknowledged with reference to the depth of ideas and how much the idea and the underlying human need have been made explicit in order to be understood.

In sum, the benefits of collaboration with potential users are rooted within the enriched collateral benefits drawn from compensation and the subsequent process with an experienced person. Support, critique and improvisation substantiate, extract and formalize the unique qualities of the emotional and experiential insights. It is a naturally emerging two-step process of finding a real-life problem, before finding solutions within a series of single insights and actions that lead to something meaningful. The often referenced merit of identifying upcoming needs when collaborating with (potential) users (e.g. Danneels, 2002; Mascitelli, 2000) is thereby conceptualized with its embodiment in communication.

The common and widely acknowledged theories from sociocognitve psychology are reflected in the findings, but cannot conclusively explain the merits of potential user collaboration. The findings show a strong relation to the sociocultural mindset of creativity and its focus on emergent properties (Sawyer, 2005). Organizational research has acknowledged the relation between improvisation and organizational learning in new product development (Crossan, 1998; Cunha et al., 1999; Kyriakopoulos, 2011; Moorman & Miner, 1998). Altering an idea on the go is an *artifactual production*, since it adds or subtracts components or features from the initial idea. Redefining a critique into an *interpretive production*, as it changes the perspective on inherent components or features of an idea (Miner et al., 2001). The virtue of potential user collaboration can be compared to the playfulness and lightheartedness of children to improvise in creative activities. Children start improvising more naturally than adults (Baker-Sennett et al., 1992; Bearison & Dorval, 2002). They accept that not all properties of an output need to be elaborated before it can be shared, tested and also radically changed. Furthermore, their general lack of knowledge and experience urges them to communicate and collaborate (Kratzer & Lettl, 2008). However, "improvisation is not inherently good or bad" (Vera & Crossan, 2001, p. 204) for the outcome as it can be creative but also chaotic. The antecedents of improvisation, namely real time exchange (Vera & Crossan, 2001), the enrichment of the information by finding real-life problems before contriving solutions, the intensive sense-making (meaning giving) and the rise of a collective cognition, these taken together allow for the improvisation to result predominantly in an original output. The nature of the problem (ill-definition) influences the process as an intervening condition. The absence of an idealized end-state - no right or wrong, only good or bad ideas - encourages the collaboration to improvise.

1.4 Conclusion and Implications

The overall goal of this dissertation is to develop a comprehensive picture of the merits of collaboration with potential users, who are at the peripheries of markets. Though literature on innovation management has abundantly referenced the need to incorporate these distant markets as a means to innovate and remain competitive, there has not been a systematic inquiry into gains and drawbacks on a micro process level (Greer & Lei, 2012; Kristensson et al., 2008). Also, potential user's suitability in terms of performance in active collaboration has not been validated from an individual perspective. There are no guidelines for organizations on how to classify potential users, how to collaborate with, and what to expect from them. Many organizations find it difficult to deviate from their familiar paths of exploration and rely solely on their current users (Bond & Houston, 2003; Prahadad & Ramaswamy, 2000).

Building on the paradigms of mixed method research (Johnson et al., 2007; Teddlie & Tashakkori, 2011) and triangulation (Denzin, 1989), the articles here apply both a quantitative (validation) and qualitative (inquiry) experimental design with a focus on the individual, the collaborative process and its emergences and the output of the process.

Article I deals with the definition of a potential user and how collaboration with potential users is embedded in research on innovation management and creativity. In the course of the research that went into this article, significant research gaps were identified. The article forms the conceptual basis on which the experimental design was operationalized.

Article II seeks to validate the relationship between the prior experience of a user, the cognitive distance between collaborating users in dyads, and the discontinuity of contributed ideas. Differentiating the expert's evaluations of the ideas into market and technological discontinuity reveals particular constituents of the ideas and how they relate to organizational competences. Furthermore, limits to the user involvement notion in general, and to the idea of involving potential users in particular, are disclosed and discussed.

Article III is an in-depth inquiry into the uniqueness of individual contributions by potential users, how these influence the collaborative process in different dyadic compositions and how they are influenced by what emerges within the process. Not neglecting the individual level of analysis, this article also takes a sociocultural perspective on the collaborative process and the intricate relations between the individual and the context.

The general finding from this comprehensive validation and inquiry is that potential users are valuable partners in collaborative exploration and that their virtue arises from a collaborative effort with experienced individuals. In summary, the following conclusions can be drawn from the dissertation and its three articles with regards to the research questions:

The constituents and impact of the merit of collaboration with potential users

1. The ideas of potential users from heterogeneous dyads are more discontinuous to established market characteristics than those of current users. They are characterized in particular by a degree of novelty with regard to dominant designs and an element of surprise is embedded in the idea. These irregularities deviate from common expectations.

2. The high discontinuity to established market characteristics of potential users' ideas does not affect the level of technological change required for implementation. The collaboration with potential users therefore suggests ways to leverage the technological competences of an organization. It extracts new value-of-use by de- and relinking technology to new social contexts.

3. In user collaboration, cognitive flexibility or cognitive distance, taken alone, are not sufficient to reach above average results. Potential users from homogeneous dyads did not perform better than current users. Moreover, current users from heterogeneous dyads did not benefit from cognitive distance in the same way as the potential users. This asymmetry has its origins in the nature of the shared knowledge and personal agendas.

The embodiment of the merit in the collaborative process

4. On a general account, the contributions, stimuli and ideas of potential users are meaningful and enriched by social, cultural and emotional values. It is partly a collateral benefit of their behavior to compensate their experience gap. Deliberate time is spent on finding a meaningful problem, before contemplating solutions. Potential users' have a natural approach towards problem-finding creativity.

5. The collateral benefit of compensation arises from the observed problem-finding approaches and behavior of *declarative simplification, context/emotion-related critical incidents, personal transfer (interdomain)* and the *personal agenda* which pervade the collaborative communication.

6. The enrichment of the idea impulses and the relation to social, cultural and emotional values is amplified when potential users collaborate with experienced individuals through thorough exchange (*sense-making*) and (un-) learning (*assimilation*).

7. The merit of collaboration with potential users has a procedural trait. The deliberate debate facilitates the depth of ideas and its tangibility. The willingness of potential users to improvise leads to a series of small ad-hoc changes (circumvention or redefinition) of the initial idea that together foster transformation. This process partly attenuates the 'mystic' conception of a sudden insight arising through cognitive flexibility. Different objectives and agendas between potential and current users create a deliberate debate which catalyzes the process to connect the known with the new and make it understandable. This is a collaborative effort which not only brings up meaningful preliminary ideas, but also shapes them into promising opportunities.

Antecedents and conditions of the merit

8. The ill-definition of the creative task encourages the collaborative partners to improvise as there is no right or wrong answer to the task.

9. Cognitive distance between the collaborating partners and the activity of contextual framing are catalysts towards meaningful ideas stimulated by sense-making, assimilation and the prevention of mere anticipation.

10. Face-to-face interaction increases the attendance to articulate delicate and enriched information (e.g. emotional involvement) and thereby enables the parallel expression of critique perceived as challenge and mutual guidance between potential and current users.

The differences in the evaluation of the output, as analyzed in article II, emerged from differences in the processes that dynamically evolved with different levels of predictability as discovered in article III. The merit of potential user collaboration, therefore, cannot be explained by an individualistic focus only. The notion of enrichment seems also to be reflected in the study of article II, in which the ideas contributed by potential users from heterogeneous dyads - though considered more original and surprising - did not require considerable technological leaps. The output was therefore not particularly novel from a technological perspective, but was indeed so in terms of its cultural relevance. The salient ideas of potential users from heterogeneous dyads originate in the elaborate social process that emerges with the current user.

This dissertation contributes to existing research on multiple levels. Firstly, it provides a precise reconstruction of the processes with which users contribute valuable information in collaboration. Although user involvement in innovation has been a focus of study for over two decades (e.g. Bharadwaj et al., 2012; Foxall, 1989; Kristensson et al., 2004; Lettl et al., 2006; Salomo et al., 2003; von Hippel, 1986), research on the micro processes to acquire information in collaboration with (potential) users have been rare and inconclusive (Greer & Lei, 2012; Kristensson et al., 2008). This dissertation infuses concepts and procedural traits to the often stated, but conceptually undefined, goal of uncovering unfulfilled needs in collaboration with users. It focuses on the potential user at the edges of markets, the virtues of which have never been investigated systematically. The findings draw a comprehensive picture of the merit of collaboration with potential users and its embodiment in collaborative communication. They therefore mitigate the selection problem of which users to involve and how to collaborate in organizational practice when an innovation project is explicitly expected to foster transformation (Kristensson & Magnusson, 2010; Nijssen et al., 2012). The results do not oppose but strongly complement the seminal lead user concept (von Hippel, 1986). The integration of lead users is acknowledged to be beneficial not only in environmental search processes, but also in the exploitation phase of the innovation process. In the former, however, creativity research informs us that it is not always the deeply knowledgeable and interwoven person who presents the ability to radically change a domain (e.g. Duncker, 1945; Rubenson & Runco, 1995; Weisberg, 1999). This is particularly relevant when the goal is to redefine the boundaries of a domain through explorative learning (Lettl, 2007).

Secondly, the present work contributes to research on the front-end of new product development, where preliminary stimuli and ideas are introduced to the innovative activities of organizations (e.g. Kim & Wilemon, 2002; Reid & Brentani, 2004). Organizations do not necessarily suffer from a lack of good ideas, but rather from a lack of insight into how to process and shape preliminary ideas into promising opportunities (Florén & Frishammar, 2012). Collaboration between potential and current users infuses the ideas with meaning and contextual information. Organizations can thus learn how to take advantage of their technical competences in new terrains and make better choices in the front end process. As a consequence, abilities to allow improvisation and endure ambiguity need to be strengthened.

Thirdly, current research on creativity in exploration has applied theories from sociocognitive psychology, while the sociocultural paradigm has received little attention. Though sociocognitive theories of creativity have been validated in the context of collaboration with potential users, they cannot conclusively explain the merits. Analysis of the collaborative processes, as presented here, has shown that in particular the unintended and emergent properties of the collaboration reflect explanatory approaches to the salient output of collaboration with potential users. This is a first attempt to apply the sociocultural mindset to user research. Indeed, new concepts, in particular applying to the explorative context in collaboration between users, have been established and elaborated. The results are therefore also a call to researchers in innovation management to give the contextual approaches to creativity more attention.

Fourthly, the work shows antecedents, influences and outcomes of improvisation from organizational learning theories (e.g. Kamoche, et al., 2003; Kyriakopoulos, 2011;

Moorman & Miner, 1998), analogical thinking and transfer (e.g. Dahl & Moreau, 2002; Dunbar, 1997) and declarative simplification strategies (Gassmann & Zeschky, 2008; Ward, 2004) in the participative product innovation process with users. Novelty was created by (1) circumventing challenges by altering constituents of the idea in real-time and (2) redefining the inherent meaning of the critique. This process can be compared to cycles of composition and execution from improvisational learning (Moorman & Miner, 1998).

Fifthly, the research contributes a unique definition and classification scheme for potential users, a validation of their suitability and an in-depth inquiry into the literature that deals with organizational strategies towards discontinuous innovations and new market opportunities as well as their requests to involve distant environments into the development of new products and services (e.g. Christensen, 2006; Danneels, 2007; Day, 1999; Faems et al., 2005; Gilbert, 2003; Lynn et al, 1996; Markides, 2006; Slater & Naver, 1998).

Some of the limitations of this dissertation and paths for future research should be mentioned. The overall research design encompasses a quantitative method as complementary to an extensive qualitative research (Miles & Huberman, 1994). For the quantitative study, the sample size is admittedly small with 68 individuals when it comes to the simple effect comparisons of the varying factorial levels. Though block designs generally reduce error variance (Liebert & Liebert, 1995), increasing the sample size would yield a more robust result with regards to chance factors. Furthermore, using a full design with control groups of current and potential users who did not collaborate with someone before would strengthen the quantitative validation. Changes in the setting of the qualitative study suggest promising research paths. The data resulted, amongst other sources, from the ethnography of communication (Hymes, 1974; Saville-Troike, 2003). Communication within heterogeneous dyads, in particular, is influenced by the complexity of the domain. In these microcells of collaborative dyads and groups, communication becomes especially important (Sawyer, 2005), as it constructs the context and defines the working dynamics. Similar studies in various domains with varying complexity and a focus on communication are a promising path for future research.

Furthermore, in organizational practice, user involvement does not always require the user to be involved actively. The study and comparison of potential users in the continuum between passive and active participation therefore leaves room for various interesting study designs. The idea discussions were conducted between two users with similar or distinct preconditions. It would certainly be fruitful to extend the research to the corporate level, where a member of an organization would discuss ideas with both potential and current users. For one, their expertise in identifying and synthesizing environmental stimuli and sticky information is likely to mitigate the asymmetric knowledge exchange. Secondly, they would include different realms of knowledge resulting from their ability, for instance, to recognize the kind of ideas that fit well within the organization or those that have been rewarding in the past (Sternberg, 1998). Knowledge of this kind is bound to affect the creative problem-solving process. Moreover, contrasting potential and lead users' contributions and impact seems insightful, as involving potential users might not always be the dominant strategy to develop new markets in collaboration with users.

The results of this dissertation also comprise managerial implications, as they can challenge the status-quo of user involvement in organizational practice. Organizations that facilitate creative collaboration activities have long followed the postulate of choosing quantity over quality of ideas. The underlying statistical reasoning behind this is that the more ideas that have been thought of at the end of the day, the higher the probability that the ideas cover a wide spectrum and that at least a few will be new and surprising (Goldenberg et al., 1999; Simonton, 1977). However, when collaborating with potential users, a shift towards fewer ideas with more depth and elaboration can be shown to result in a more effectively steered exploration. Collaborations with potential users should be given focus and debate, not a climate of irrelevance to any boundaries. Furthermore, research on group creativity recommends deferring any critique in idea generation (Osborn, 1963). This view ignores, however, two essential qualities that critique, if constructively reported, can contribute. First, critique transports knowledge and therefore stimulates mutual learning and unlearning in collaboration. Spending time on fewer thoughts and giving rise to constructive critique can help one to understand the unique qualities of the contexts (Visser et al., 2007). Instead of getting more of the same, a relation to new and useful contexts is highlighted which makes the social context more understandable. Second, if perceived as a challenge, critique once again fuels the process of imagination and fosters deviation from the norm through iterations between improvisation and planning (Dunbar, 1997). The role of the key individual of the organization shifts to the facilitator. This person acquires the task of identifying meaning, insight and inspiration within emotionally and situationally bound communication. Potential user collaboration is a real-time and short-time learning process that requires cycles and iterations. The stimuli for new ideas is the germ of successful innovation (Henard & Szymanski, 2001) and a well-managed exploration with potential users can teach an organization which ways are meaningful and why.

References

Abernathy, W. J., & Clark, K. B. 1985. Innovation: Mapping the Winds of Creative Destruction. *Research Policy,* 14: 3-22.

Alba, J. W., & Hutchinson, W. J. 1987. Dimensions of Consumer Expertise. *Journal of Consumer Research,* 13(4): 411–454.

Amabile, T. M. 1996. *Creativity in Context: Update to the Social Psychology of Creativity.* Boulder, CO: Westview Press.

Anthony, S. D. 2009. Major League Innovation. *Harvard Business Review*, 87(10): 51-54.

Archer, M. 1995. *Realist Social Theory: The Morphogenetic Approach.* New York: Cambridge University Press.

Arnold, T. J., Fang, E., & Palmatier, R. W. 2011. The Effects of Customer Acquisition and Retention Orientations on a Firm's Radical and Incremental Innovation Performance. *Journal of the Academy of Marketing Science*, 39(2): 234-251.

Audia, P. G., & Goncalo, J. A. 2007. Past Success and Creativity over Time: A Study of Inventors in the Hard Disk Drive Industry. *Management Science,* 53(1): 1-15.

Bailey, R. A. 2008. *Design of Comparative Experiments.* Cambridge Series in Statistical and Probabilistic Mathematics, Queen Mary, University of London.

Baker-Sennett, J., Matusov, E., & Rogoff, B. 1992. Sociocultural Processes of Creative Planning in Children's Playcrafting. In P. Light & G. Butterworth (Eds.), *Context and Cognition: Ways of Learning and Knowing:* 93-114. New York: Harvester-Wheatsheaf.

Baldwin, C., Hienerth, C., & von Hippel, E. 2006. How User Innovations Become Commercial Products: A Theoretical Investigation and Case Study. *Research Policy,* 35: 1291–1313.

Bearison, D. J., & Dorval, B. 2002. *Collaborative Cognition: Children Negotiating Ways of Knowing.* Westport, CT: Ablex Publishing.

Bharadwaj, N., Nevin, J. R., & Wallman, J. P. 2012. Explicating Hearing the Voice of the Customer as a Manifestation of Customer Focus and Assessing its Consequences. *Journal of Product Innovation Management*, forthcoming.

Bogers, M., Afuah, A., & Bastian, B. 2010. Users as Innovators: A Review, Critique, and Future Research Directions. *Journal of Management*, 36: 857–875.

Bond, E., & Houston, M. 2003. Barriers to Matching New Technologies and Market Opportunities in Established Firms. *Journal of Product Innovation Management*, 20(2): 120-135.

Bonner, J. M., & Walker Jr., O. C. 2004. Selecting Influential Business-to-Business Customers in New Product Development: Relational Embeddedness and Knowledge Heterogeneity Considerations. *Journal of Product Innovation Management*, 21: 155-169.

Brucks, M. 1985. The Effects of Product Class Knowledge on Information Search Behavior. *Journal of Consumer Research*, 12(1): 1–16.

Bruner, J. 1990. *Acts of Meaning.* Cambridge, MA: Harvard University Press.

Callahan, J., & Lasry, E. 2004. The Importance of Customer Input in the Development of Very New Products. *R&D Management*, 34(2):107–120.

Campbell, D. T. 1960. Blind Variation and Selective Retention in Creative Thought as in other Knowledge Processes. *Psychological Review*, 67: 380–400.

Chandy, R. K., & Tellis, G. J. 1998. Organizing for Radical Product Innovation: The Overlooked Role of Willingness to Cannibalize. *Journal of Marketing Research*, 35(4): 474–487.

Chandy, R. K., & Tellis, G. J. 2000. The Incumbent's Curse? Incumbency, Size, and Radical Product Innovation. *Journal of Marketing*, 64(3): 1-17.

Christensen, C. M. 1997. *The Innovator's Dilemma.* Boston: Harvard Business School Press.

Christensen, C. M. 2006. The Ongoing Process of Building a Theory of Disruption. *Journal of Product Innovation Management*, 23: 39-55.

Christensen C. M., & Bower, J. L. 1996. Customer Power, Strategic Investment, and the Failure of Leading Firms. *Strategic Management Journal*, 17(3): 197–218.

Clark, V. L. P., Creswell, J. W., Green, D. O., & Shope, R. J. 2008. Mixing Quantitative and Qualitative Approaches: An Introduction to Emergent Mixed Methods Research. In S. N. Hesse-Biber & P. Leavy (Eds.), *Handbook of Emergent Methods:* 363-388. New York: Guilford.

Cole, M. 1996. *Cultural Psychology: A Once and Future Discipline.* Cambridge, MA: Belknap Press.

Cook, T. D., & Shadish, W. R. 1994. Social Experiments: Some developments over the past 15 years. *Annual Review of Psychology*, 45: 545-580.

Crossan, M. M. 1998. Improvisation in Action. *Organization Science*, 9(5): 593-599.

Cunha, M. P., Cunha, V. J., & Kamoche, K. 1999. Organizational Improvisation: What, When, How and Why. *International Journal of Management Reviews*, 1: 299–341.

Dahl, D. W., & Moreau, P. 2002. The Influence and Value of Analogic Thinking During New Product Ideation. *Journal of Marketing Research*, 39: 47-60.

Danneels, E. 2002. The Dynamics of Product Innovation and Firm Competences. Strategic Management Journal, 23: 1095-1121.

Danneels, E. 2003. Tight-Loose Coupling with Customers: The Enactment of Customer Orientation. *Strategic Management Journal*, 24: 559-576.

Danneels, E. 2007. The Process of Technological Competence Leveraging. *Strategic Management Journal*, 28(5): 511-533.

Day, G. S. 1999. Misconceptions about Market Orientation. *Journal of Market Focused Management*, 4: 5-16.

Denzin, N. K. 1989. *The Research Act: A Theoretical Introduction to Sociological Methods*. Third edition, New York.

Denzin, N. K., & Lincoln, Y. S. 2011. Introduction: The Discipline and Practice of Qualitative Research. In N. K. Denzin & Y. S. Lincoln (Eds.), *The Sage Handbook of Qualitative Research (4th ed.):* 1-19. Sage.

Dougherty, D. 1992. A Practice-Centered Model of Organizational Renewal through Product Innovation. *Strategic Management Journal*, 13: 77–92.

Dunbar, K. 1997. How Scientists Think: On-Line Creativity and Conceptual Change in Science. In T. B. Ward, S. M. Smith & J. Vaid (Eds.), *Conceptual structures and processes: Emergence, Discovery, and Change:* 461-493: Washington D.C: American Psychological Association Press.

Duncker, K. 1945. On Problem-solving. *Psychological Monographs*, 58(5): whole no. 270.

Edmondson, A. 1999. Psychological Safety and Learning Behavior in Work Teams. *Administrative Science Quarterly*, 44(2): 350-383.

Faems, D., van Looy, B., & Debackere, K. 2005. Interorganizational Collaboration and Innovation: Toward a Portfolio Approach. *Journal of Product Innovation Management*, 22: 238–250.

Fernández-Cárdenas, J. M. 2008. The Situated Aspect of Creativity in Communicative Events: How do Children Design Web Pages Together? *Thinking Skills and Creativity*, 3: 203-216.

Florén, H., & Frishammar, J. 2012. From Preliminary Ideas to Corroborated Product Definitions: Managing the Front End of New Product Development. *California Management Review*, 54(4): 20-43.

Foxall, G. R. 1989. User Initiated Product Innovations. *Industrial Marketing Management*, 18(2): 95–104.

Fuchs, C., & Schreier, M. 2011. Customer Empowerment in New Product Development. *Journal of Product Innovation Management*, 28(1): 17-32.

Garcia, R., & Calantone, R. 2002. A Critical Look at Technological Innovation Typology and Innovativeness Terminology: A Literature Review. *Journal of Product Innovation Management*, 19(2): 110–132.

Gardner, H. 1985. *The Mind's New Science: A History of the Cognitive Revolution.* New York: Basic Books.

Gassmann, O., & Zeschky, M. 2008. Opening up the Solution Space: The Role of Analogical Thinking for Breakthrough Product Innovation. *Creativity and Innovation Management,* 17(2): 97-106.

Gilbert, C. 2003. The Disruption Opportunity. *Sloan Management Review,* 44(4): 27-32.

Given, L. M. (Ed.) 2008. *The Sage Encyclopedia of Qualitative Research Methods,* vol. 2: 697-698. Sage: Thousand Oaks, CA.

Glaser, B. G., & Strauss, A. L. 1967. *Discovery of Grounded Theory: Strategies for Qualitative Research.* Chicago: Aldine.

Glaveanu, V. P. 2011. How are We Creative Together? Comparing Sociocognitive and Sociocultural answers. *Theory and Psychology,* 21(4): 473-492.

Goldenberg, J., Lehmann, D. R., & Mazursky, D. 2001. The Idea Itself and the Circumstances of Its Emergence as Predictors of New Product Success. *Management Science,* 47: 69-84.

Goldenberg, J., Mazursky, D., & Solomon, S. 1999. Toward Identifying the Inventive Templates of New Products: A Channeled Ideation Approach. *Journal of Marketing Research,* 36, 200–210.

Govindarajan, V., Kopalle, P. K., & Danneels, E. 2011. The Effects of Mainstream and Emerging Customer Orientations on Radical and Disruptive Innovations. *Journal of Product Innovation Management,* 28: 121-132.

Greene, J. C. 2007. *Mixing Methods in Social Inquiry.* San Francisco: Jossey-Bass.

Greer, C. R., & Lei, D. 2012. Collaborative Innovation with Customers: A Review of The Literature and Suggestions for Future Research. *International Journal of Management Reviews,* 14(1): 63–84.

Gruner, K., & Homburg, C. 2000. Does Customer Interaction Enhance New Product Success? *Journal of Business Research,* 49(1): 1-14.

Guilford, J. P. 1968. *Creativity, Intelligence, and their Educational Implications.* San Diego, CA.

Hamel, G., & Prahalad, C. K. 1991. Corporate Imagination and Expeditionary Marketing. *Harvard Business Review,* 69(4): 81–92.

Hang, C., Chen, J., & Subramian A. M. 2010. Developing Disruptive Products for Emerging Economies: Lessons from Asian cases. *Research Technology Management:* 21-26.

Hauser, J. R., Tellis, G. J., & Griffin, A. 2006. Research on Innovation: A Review and Agenda for Marketing Science. *Marketing Science,* 25(6): 687-717.

Henard, D. H., & Szymanski, D. M. 2001. Why Some New Products are more Successful than Others. *Journal of Marketing Research,* 38(3): 362-375.

Henderson, R. 2006. The Innovator's Dilemma as a Problem of Organizational Competence. *Journal of Product Innovation Management,* 23: 5-11.

Hennessey, B. A. & Amabile, T. 2010. Creativity. *Annual Review of Psychology,* 61: 569–598.

Hoffman, D. L., Kopalle, P. K., & Novak, T. P. 2010. The 'Right' Consumers for Better Concepts: Identifying and Using Consumers High in Emergent Nature to Further Develop New Product Concepts. *Journal of Marketing Research,* 47: 854-865.

Hutchins, E. 1995. *Cognition in the Wild.* Cambridge, MA: MIT Press.

Hymes, D. 1974. *Foundations in Sociolinguistics.* London: Tavistock Publications Limited.

Im, S., Bayus, B. L., & Mason, C. H. 2003. An Empirical Study of Innate Consumer Innovativeness, Personal Characteristics, and New-Product Adoption Behavior. *Journal of the Academy of Marketing Science,* 31: 61–73.

Jackson, P., & Messick, S. 1965. The Person, the Product and the Response: Conceptual Problems in the Assessment of Creativity. *Journal of Personality,* 33: 309-329.

Johnson, R. B., Onwuegbuzie, A. J., & Turner, L. A. 2007. Toward a Definition of Mixed Methods Research. *Journal of Mixed Methods Research,* 1(2): 112-133.

Kamoche, K, Pina e Cunha, M., & Vieira da Cunha, J. 2003. Towards a Theory of Organizational Improvisation: Looking Beyond the Jazz Metaphor. *Journal of Management Studies,* 40(8): 2023-2051.

Kim, J., & Wilemon, D. 2002. Focusing the Fuzzy Front-end in New Product Development. *R&D Management,* 32(4): 269-279.

Kirton, M. 1976. Adaptors and Innovators: A Description and Measure. *Journal of Applied Psychology,* 61(5): 622–29.

Kleining, G. 1995. Das Qualitative Experiment. In U. Flick, E. v. Kardorff, H. Keupp, L. v. Rosenstiel, & S. Wolff (Eds.), *Handbuch Qualitative Sozialforschung:* 263-266. Weinheim.

Kratzer, J., & Lettl, C. 2008. A Social Network Perspective of Lead Users and Creativity: An Empirical Study among Children. *Creativity and Innovation Management,* 17(1): 26-36.

Kristensson, P., Gustafsson, A., & Archer, T. 2004. Harnessing the Creative Potential among Users. *Journal of Product Innovation Management,* 21: 4-14.

Kristensson, P., Magnusson, P. R. 2010. Tuning Users' Innovativeness During Ideation. *Creativity and Innovation Management,* 19(2): 147-159.

Kristensson, P., Matthing, J., & Johansson, N. 2008. Key Strategies for the Successful Involvement of Customers in the Co-Creation of New Technology-Based Services. *International Journal of Service Industry Management,* 19(3-4): 474-491.

Kock, A., Gemünden, H. G., Salomo, S., & Schultz, C. 2011. The Mixed Blessings of Technological Innovativeness for the Commercial Success of New Products. *Journal of Product Innovation Management,* 28(S1): 28-43.

Kyriakopoulos, K. 2011. Improvisation in Product Innovation: The Contingent Role of Market Information Sources and Memory Types. *Organization Studies,* 32(8): 1051–1078.

Larey, T., & Paulus, P. 1999. Group Preference and Convergent Tendencies in Small Groups: A Content Analysis of Group Brainstorming Performance. *Creativity Research Journal,* 12(3): 175–184.

Lau, A. K. W., Tang, E., & Yam, R. C. M. 2010. Effects of Supplier and Customer Integration on Product Innovation and Performance: Empirical Evidence in Hong Kong Manufacturers. *Journal of Product Innovation Management*, 27(5): 761-777.

Lei, D., & Greer, C. R. 2003. The Empathetic Organization. *Organizational Dynamics*, 32: 142–164.

Lettl, C. 2007. User Involvement Competence for Radical Innovation. *Journal of Engineering and Technology Management*, 24: 53-75.

Lettl, C., Herstatt, C., & Gemünden, H. G. (2006). Users' Contributions to Radical Innovation: Evidence from Four Cases in the Field of Medical Equipment Technology. *R&D Management*, 36: 251-272.

Leonard, D., & Rayport, J. F. 1997. Spark Innovation through Empathic Design. *Harvard Business Review*, 75: 102-113.

Leonard, D., & Sensiper, S. 1998. The Role of Tacit Knowledge in Group Innovation. *California Management Review*, 40: 112–132.

Liebert, R. M., & Liebert, L. L. 1995. *Science and Behavior: An Introduction to Methods of Psychological Research*. Englewood Cliffs, NJ: Prentice Hall.

Litchfield, R. C. 2008. Brainstorming Reconsidered: A Goal-Based View. *Academy of Management Review*, 33(3): 649–668.

Lynn, G. S., Morone, J. G., & Paulson, A. S. 1996. Marketing and Discontinuous Innovation: the Probe and Learn Process. *California Management Review*, 38(3): 8-37.

Markides, C. 2006. Disruptive Innovation: In Need of Better Theory. Journal of Product Innovation Management, 23: 19-25.

Mascitelli, R. 2000. From Experience: Harnessing Tacit Knowledge to Achieve Breakthrough Innovation. *Journal of Product and Innovation Management*, 17: 179-193.

McGraw, K. O., & Wong, S. P 1996. Forming Inferences about some Intraclass Correlation Coefficients. *Psychological Methods*, 1: 651–655.

Miles, M. B., & Huberman, A. M. 1994. *Qualitative Data Analysis: An Expanded Sourcebook*. Thousand Oaks: Sage Publications.

Miller, G. A. 2003. The Cognitive Revolution: A Historical Perspective. *Trends in Cognitive Sciences*, 7(3): 141-144.

Miner, A. S., Bassoff, P., & Moorman, C. 2001. Organizational Improvisation and Learning: A Field Study. *Administrative Science Quarterly*, 46(2): 304-337.

Moorman, C., & Miner, A. S. (1998). Organizational Improvisation and Organizational Memory. *Academy of Management Review*, 23: 698–723.

Moran, S., & John-Steiner, V. (2004). How Collaboration in Creative Work Impacts Identity and Motivation. In D. Miell & K. Littleton (Eds.), *Collaborative Creativity: Contemporary Perspectives:* 11–25. London, UK: Free Association Books.

Narver, J. C., Slater, S. F., & MacLachlan, D. L. 2004. Responsive and Proactive Market Orientation and New-product Success. *Journal of Product Innovation Management*, 21(5): 334-47.

Newell, A., Shaw, J. C., & Simon, H. A. 1958. Elements of a Theory of Human Problem-solving. *Psychological Review*, 23: 342-343.

Nijssen, E. J., Hillebrand, B, de Jong, J. P. J., & Kemp, R. G. M. 2012. Strategic Value Assessment and Explorative Learning Opportunities with Customers. *Journal of Product Innovation Management*, 29(S1): 91-102.

Nooteboom, B. 2009. *A Cognitive Theory of the Firm: Learning, Governance and Dynamic Capabilities.* Edward Elgar Publishing Ltd.

Ojanen, V., & Hallikas, J. 2009. Inter-Organizational Routines and Transformation of Customer Relationships in Collaborative Innovation. *International Journal of Technology Management*, 45: 306–322.

Osborn, A. F. 1963. *Applied Imagination: Principles and Procedures of Creative Problem-Solving.* New York, NY: Charles Scribner's Sons.

Paap, J., & Katz, R. 2004. Anticipating Disruptive Innovation. *Research Technology Management*, 47(5): 13-22.

Park, W. C., Mothersbaugh, D. L., & Feick, L. 1994. Consumer Knowledge Assessment. *Journal of Consumer Research*, 21(1): 71–82.

Paulus, P., Nakui, T., & Putman, V. 2006. Group Brainstorming and Teamwork: Some Rules for the Road to Innovation. In L. Thompson & H.-S. Choi (Eds.), *Creativity and Innovation in Organizational Teams:* 69–86. Mahwah, NJ: Lawrence Erlbaum Associates.

Paulus, P. B., & Nijstad, B. 2003. Group Creativity: An Introduction. In P. Paulus & B. Nijstad (Eds.), *Group Creativity: Innovation Through Collaboration:* 3–11: New York, NY: Oxford University Press.

Paulus, P. B., & Yang, H. C. 2000. Idea Generation in Groups: A Basis for Creativity in Organizations. *Organizational Behavior and Human Decision Processes*, 82: 76-87.

Piller, F. T., & Walcher, D. 2006. Toolkits for Idea Competitions: A Novel Method to Integrate Users in New Product Development. *R&D Management*, 36(3): 307-318.

Prahalad, C. K., & Ramaswamy, V. 2000. Co-opting Customer Competence. *Harvard Business Review*, 78(1): 82.

Prahalad, C. K., & Ramaswamy, V. 2004. Co-Creating Unique Value with Customers. *Strategy and Leadership*, 32 (3): 4–9.

Pratt, M. G. 2008. Fitting Oval Pegs Into Round Holes - Tensions in Evaluating and Publishing Qualitative Research in Top-Tier North American Journals. *Organizational Research Methods*, 11(3): 481-509.

Pratt, M. G. 2009. From the Editors: For the lack of a Boilerplate: Tips on Writing up (and Reviewing) Qualitative Research. *Academy of Management Journal*, 52(2): 856-862.

Reid, S. E., & Brentani, U. 2004. The Fuzzy Front End of New Product Development for Discontinuous Innovations: A Theoretical Model. *Journal of Product Innovation Management*, 21: 170-184.

Rogoff, B. 1995. Observing Sociocultural Activity on Three Planes: Participatory Appropriation, Guided Participation, and Apprenticeship. In J. V. Wertsch, P. del

Rio, & A. Alvarez (Eds.), *Sociocultural Studies Of Mind:* 139-164. Cambridge, UK: Cambridge University Press.

Rubenson, D. L., & Runco, M. A. 1995. The Psychoeconomic View of Creative Work in Groups and Organizations. *Creativity and Innovation Management,* 4: 232-241.

Runco, M. A., & Sakamoto, S. O. 1999. Experimental Studies of Creativity. In R. J. Sternberg (Ed.), *Handbook of Creativity:* 62-92. New York: Cambridge University Press.

Salomo, S., Steinhoff, F., & Trommsdorff, V. 2003. Customer Orientation in Innovation Projects and New Product Development Success - The Moderating Effect of Product Innovativeness. *International Journal of Technology Management,* 26: 442-463.

Samra-Fredericks, D. 2004. Talk-in-Interaction/Conversation Analysis. In C. Cassell & G. Symon (Eds.), *Essential Guide to Qualitative Methods in Organizational Research:* 215-225. London: Sage.

Saville-Troike, M. 2003. *The Ethnography of Communication.* Oxford: Blackwell Publishing.

Sawyer, R. K. 2005. *Social Emergence.* Cambridge University Press, Cambridge.

Sawyer, R. K. 2007. *Group Genius.* New York, NY: Basic Books.

Seddon, F. 2004. Empathic Creativity: The Product of Empathic Attunement. In D. Miell & K. Littleton (Eds.), *Collaborative creativity: Contemporary perspectives*: 65–78. London, UK: Free Association Books.

Shaw, B. S. 1985. The Role of the Interaction between the User and the Manufacturer in Medical Equipment Innovation. *R&D Management,* 15: 283-292.

Sherman, L. W., & Strang, H. 2004. Experimental Ethnography: The Marriage of Qualitative and Quantitative Research. *Annals of the Academy of Social and Political Science,* 595: 204-222.

Shrout, P. E., & Fleiss, J. L. 1979. Intraclass Correlations: Users in Assessing Rater Reliability. *Psychological Bulletin,* 86: 420–428.

Shweder, R. 1990. Cultural Psychology – What is it? In J. Stigler, R. Shweder, & G. Herdt (Eds.), *Cultural Psychology: Essays on Comparative Human Development*: 1–43: Cambridge, UK: Cambridge University Press.

Siever, R. 1968. Science: Observational, Experimental, Historical. *American Scientist,* 56 (1): 70-77.

Simonton, D. K. 1977. Creative Productivity, Age and Stress: A Biographical Time-series Analysis of 10 Classical Composers. *Journal of Personality and Social Psychology,* 35: 805–816.

Slater, S. F., & Mohr, J. J. 2006. Successful Development and Commercialization of Technological Innovation: Insights Based on Strategy Type. *Journal of Product Innovation Management,* 23: 26-33.

Slater, S. F., & Narver, J. C. 1998. Customer-led and Market Oriented: Let's Not Confuse the Two. *Strategic Management Journal,* 19(10): 1001–1006.

Smith, S., Gerkens, D., Shah, J., & Vargas-Hernandez, N. 2006. Empirical Studies of Creative Cognition in Idea Generation. In L. Thompson & H.-S. Choi (Eds.),

Creativity and Innovation in Organizational Teams: 3–20. Mahwah, NJ: Lawrence Erlbaum Associates.

Sonnenburg, S. 2004. Creativity in Communication: A Theoretical Framework for Collaborative Product Creation. *Creativity and Innovation Management,* 13: 254-262.

Sternberg, R. J. 1998. Cognitive Mechanisms in Human Creativity: Is Variation Blind or Sighted? *Journal of Creative Behavior,* 32: 159-176.

Sternberg, R. J., Kaufman, J. C., & Pretz, J. E. 2003. A Propulsion Model of Creative Leadership. *Leadership Quarterly,* 14: 455–473.

Steyaert, C., & Bouwen, R. 2004. Group Methods of Organizational Analysis. In C. Cassell & G. Symon (Eds.), *Essential Guide to Qualitative Methods in Organizational Research:* 140-153. London: Sage.

Strauss, A. & Corbin, J. 1996. *Grounded Theory.* Weinheim: Beltz.

Teddlie, C., & Tashakkori, A. 2011. Mixed Methods Research: Contemporary Issues in an Emerging Field. In N. K. Denzin, & Y. S. Lincoln (Eds.), *The Sage Handbook of Qualitative Research (4th ed.):* 285-299: Sage.

Teddlie, C., Tashakkori, A., & Johnson, B. 2008. Emergent Techniques in the Gathering and Analysis of Mixed Method Data. In S. N. Hesse-Biber & P. Leavy (Eds.), *Handbook of Emergent Methods:* 389-413. New York: Guilford.

Thompson, L. 2004. *Making the Team: A Guide for Managers (2nd ed.).* Upper Saddle River, NJ: Pearson Education, Inc.

Tushman, M. L. & Anderson, P. 1986. Technological Discontinuities and Organizational Environments. *Administrative Science Quarterly,* 31(3): 439-465.

Tweney, R. D. 2004. Replication and the Experimental Ethnography of Science. *Journal of Cognition and Culture,* 4: 731-758.

Tzeng, C. H. 2009. A Review of Contemporary Innovation Literature: A Schumpeterian Perspective. *Innovation: Management, Policy and Practice,* 11: 373–394.

Ulwick, A. W. 2002. Turn Customer Input into Innovation. *Harvard Business Review:* 91-97.

van der Panne, G., van Beers, C., & Kleinknecht, A. 2003. Success and Failure in Innovation: A Literature Review. *International Journal of Innovation Management,* 7: 309–338.

Vargo, S. L., & Lusch, R. F. 2004. Evolving to a New Dominant Logic for Marketing. *Journal of Marketing,* 68(1): 1-17.

Vera, D., & Crossan, M. 2005. Improvisation and Innovative Performance in Teams. *Organization Science,* 16(3): 203–224

Veryzer, R. W. 1998. Key Factors Affecting Customer Evaluation of Discontinuous New Products. *Journal of Product Innovation Management,* 15(2):136–50.

Visser, F. S., van der Lugt, R., & Stappers, P. J. 2007. Sharing User Experiences in the Product Innovation Process: Participatory Design Needs Participatory Communication. *Creativity and Innovation Management,* 16: 35–45.

von Hippel, E. 1986. Lead Users: A Source of Novel Product Concepts. *Management Science,* 32(7): 791–805.

von Hippel, E. 1994. Sticky Information and the Locus of Problem-solving: Implications for Innovation. *Management Science,* 40: 429-439.

Ward, T. B. 2004. Cognition, Creativity, and Entrepreneurship. *Journal of Business Venturing,* 19: 173–88.

Weisberg, R. W. 1999. Creativity and Knowledge: A Challenge to Theories. In R. J. Sternberg (Ed.), *Handbook of Creativity:* 226-250. New York: Cambridge University Press.

Witell, L., Kristensson, P., Gustafsson, A., & Löfgren, M. 2011. Idea Generation: Customer Co-Creation versus Traditional Market Research Techniques. *Journal of Service Management,* 22(2): 140-159.

Zaichkowsky, J. L. 1985. Measuring the Involvement Construct. *The Journal of Consumer Research,* 12: 341-352.

Zeidner, M., Matthews, G., & Roberts, R. D. 2004. Emotional Intelligence in the Workplace: A Critical Review. *Applied Psychology: An international Review,* 53(3): 371–399.

2 Co-Creation with Users at the Edges of Markets

Martin Hewing and Katharina Hölzle

Abstract In order to identify growth potential in new markets, organizations constantly need to tap new sources of knowledge, especially from existing and novel users. Studies have shown that idea impulses which emerge from less experienced users turned out to be more original. This paper seeks to improve and conceptualize the understanding of collaboration with even more distant users at the boundaries of markets in the context of exploration for discontinuous innovation. Commonalities of different terms coined throughout the innovation management literature, such as potential or emerging users, are identified and a comprehensive definition is derived. The merits expected to arise from collaboration with these users, identified from a comprehensive literature review, are aligned to literature on the degree of innovativeness and theories from creativity research. Instead of solely classifying individuals according to the dichotomy between either being a current user or a non-user, the notion of use experience, user knowledge and domain involvement will be applied to allow for a more fine grained and individually based classification of users. Finally, significant research gaps in the collaboration with potential users are drawn out which call for researchers' attention in order to mitigate the struggles of organizations in explorative activities towards discontinuity.

2.1 Introduction

In order to create and address new markets, organizations constantly need to tap new sources of knowledge, especially from existing and novel users (e.g. Prahalad & Ramaswamy, 2000; von Hippel, 1986). Hence, organizations are keen to engage users early in the process of product development in order to collect ideas, feedback, and other suggestions. It is crucial in this process to understand what to expect from different types of users. While current users might contribute ideas which relate to the current design, novel users might lead a company into new but also more uncertain territory (Chandy & Tellis, 1998; Christensen, 2006; Danneels, 2004). In recent years, there have been a few research approaches aimed at studying the integration of users who do not otherwise participate in a particular market or reside in niches (e.g. Arnold et al., 2011; Chandy & Tellis 1998; Govindarajan & Kopalle, 2006a, 2006b; Govindarajan et al., 2011). However, research into the micro processes of their involvement has remained lacking, especially when it comes to collaborative settings (Greer & Lei, 2012; Kristensson et al.,

2008). Even in creativity research in general, collaborative groups have received attention only recently, when it became apparent that creative activities involve increasingly social and collaborative processes (Montuori & Purser, 1999; Sonnenburg, 2004). To the best of our knowledge, up until now, the benefit that can be drawn from collaborating with these distant users - we will refer to them as potential users from now on - is a rather vague concept in the literature on innovation. So too is the understanding of how their contributions to the idea finding phase differ from those of current users.

Before empirical analysis can be conducted, a common definition of a potential user is needed that can be operationalized. Furthermore, the notion of collaborating with potential users requires some theoretical base. The objective of this paper, therefore, is to advance research by (1) developing a comprehensive definition of the potential user, (2) identifying the merits of potential user collaboration assumed to emerge by researchers in innovation management and aligning these with theories from creativity research, (3) identifying schemes to classify users into potential or current users as referred to in available literature, (4) suggesting a classification scheme based on the derived definition using validated constructs from consumer research and (5) deriving significant research gaps that need to be tackled in order to improve the understanding of potential user collaboration.

2.2 Theoretical Perspectives and Definitions

User integration in innovation activities has been a well-regarded field of study in the last two decades (Bharadwaj et al., 2012; Foxall, 1989; Kristensson et al., 2004; Lettl et al. 2006; von Hippel, 1986). Innovations inspired by users rest upon inherent and upcoming needs and are thereby acknowledged to have a higher success probability in markets (von Hippel, 1986). The lead user concept goes even further. Lead users are not only a source of knowledge, but play a more active part in the actual conception and design of innovations (Lüthje, 2004). They are defined as people within a domain who gain relatively high benefits from solutions to their needs and therefore innovate. Lead users anticipate relevant and novel trends in a domain before many other users do (von Hippel, 1986). Their impact on innovation is thought of as being more radical, since they are deeply involved with the domain. Research on creativity, however, suggests that it is not always the knowledgeable or experienced people who have the ability to radically change a domain (e.g. Duncker, 1945; Rubenson & Runco, 1995; Weisberg, 1999). This especially applies, when the goal is to transform the boundaries of those domains. Despite acknowledgement of the benefits of user involvement, the question of which users to involve in order to foster transformation has been neglected and remains poorly understood (Kristensson & Magnusson, 2010). A comprehensive literature review shows that particularly in the context of discontinuous innovations, researchers in innovation management have recently advocated collaboration with potential users - people who are not currently served by the organization or who stay in niche markets (e.g. Christensen, 2006; Danneels, 2003; Day, 1999; Fuchs & Schreier, 2011; Gilbert, 2003; Govindarajan et al., 2011; Lau et al., 2010). But what makes their contribution in explorative search processes unique and significant compared to current users? Before an inquiry into their merit can be carried out effectively, a comprehensive definition is needed and expected merits related to existing theories on creative problem-solving in order to make use of existing theory and knowledge. In the following, we will analyze which definitions of potential users reside in the literature, how their merit in the early phase of innovation is described, how their expected merits relate to definitions and degrees of innovativeness and how users are actually classified in empirical studies. Table 2.1 shows the approach to the comprehensive literature review on which this analysis is based on. Following Kristensson et al. 2008, the terms user and customer are used interchangeably.

Table 2.1: Approach to Literature Review

Focus	Journals on innovation, organizational behavior, management, strategy, marketing & technology management
Database	EBSCO Business; ISI Web of Knowledge
Search String	"potential customer*" OR "prospective customer*" OR non-customer* OR "future customer*" OR "emerging customer*" OR "new customer*" OR "distant customer*" OR "potential user*" OR "prospective user*" OR non-user* OR "future user*" OR "emerging user*" OR "new user*" OR "distant user*" AND innovation* OR exploration* OR "organizational learning*" OR "idea generation*" OR ideation* OR co-creation*
Items	Title, abstract and keywords to select relevant articles
Additional Search	Back- & forward citations; manual searches

2.2.1 Defining the Potential User

Owing to the increased popularity of involving distant market informants in early innovation activities, plenty of synonyms for users who are not current users have arisen. One can find references to non-users (e.g. Christensen, 2006; Christensen & Bower, 1996; Danneels, 2003), potential users (e.g. Bonner & Walker, 2004; Fuchs & Schreier, 2011; Shaw, 1985), new users (e.g. Arnold et al., 2011; Danneels, 2002; Henderson, 2006), emerging users (e.g. Day, 1999; Govindarajan et al., 2011; Hoffmann et al., 2010; Slater & Mohr, 2006), prospective users (Anthony, 2009; Danneels, 2003) and future users (Bond & Houston, 2003; Chandy & Tellis, 1998). Table 2 lists relevant articles and includes examples expressing the benefits of collaboration with these users as identified throughout the review. Across studies, but also within one and the same article, the terms are used interchangeably and a statement defining the term is most often omitted. If commonalities stand out, they relate to two notions. First, the terms have in common that they refer to distant characters who do not (yet) use a product or service or reside in an unaddressed niche of the market. These people have not experienced any of the domain's products or services first hand. Second, the actual non-usage might be motivated by factors, such as a lack of a need or interest, or a lack of certain product characteristics in dominant designs. Barriers may arise from a dearth of necessary skills to use a product or service, the required means, e.g. financial, the time or access to a product or service (Christensen, 2006; Govindarajan & Kopalle, 2006a; Hang et al., 2010). This implies that these people might have an unfulfilled need or at least a positive attitude towards the domain as such. Taken together, these notions rule out occasional users who have actively decided not to be part of a market or who use a product or service with low engagement. Potential users are users who currently do not constitute a focus point for the organization's business, but have the potential to become more important in the near-by future (Govindarajan et al., 2011). Similar to Hauser et al. (2006), we define potential users *as people, who do not yet use a product or service, but who hold a positive tenor and interest towards the domain.*

Collaboration with these users is therefore not a process of blind trial and error with the vast population of non-users, but an exploration within the strategic realm of the innovating organization (Lynn et al., 1996).

2.2.2   The Merits of Collaboration with Potential Users

The merit of collaboration with potential users is mostly implied in the converse argument that "focusing exclusively on existing clientele may lead the firm to ignore potential customers, and thus miss market opportunities" (Danneels, 2003, p. 572). This statement adheres to two characteristics defining current users and their contributions in value co-creation. The first characteristic relates to the tendency of current users to want to uphold internal consistency and continuity. Current users generally use a product or service to satisfy an established need and solve a problem. Therefore, current users tend to have little or no interest in deviating too far from established dominant designs and are less inclined to dispense with highly valued performance dimensions. They want to maximize their value (Bogers et al., 2010). The behavior of the current user is described in consistency theories and their basic assumption that individuals have a need for consistency between attitudes and behaviors (e.g. Festinger, 1957; Rosenberg, 1956). A current user follows these personal objectives to prevent changes in valued product components when asked to imagine new and discontinuous products and services. Some social theories state that social properties can only change if their constituting individuals change (Sawyer, 2005). If an organization and its current users are seen as a social system, then it requires a change in behavior and perspectives on the part of current users in order to achieve transformation in dominant designs. Current users thereby exert dominance over the direction innovative activities might take towards more incremental paths (Fischer & Reuber, 2004). The second characteristic relates to the notion of functional fixedness in experienced people (e.g. Duncker, 1945). Domain-experience can obstruct the path to radical changes in the very same domain, since it requires a cognitive effort to freely shed established, habitual ways of doing things within the domain. This is a restriction that also applies to lead users (Lettl, 2007) and might explain why lead user status is not always beneficial for explorative learning (Nijssen et al., 2012). These two characteristics define the potential user only by stating what he or she is not. Potential users' imagination is thus acknowledged to be flexible (Rubenson & Runco, 1995), as it is not constrained by experience from interaction with the products or services of a domain. Potential users did not develop a clear expectation about how their needs shall be addressed.

Table 2.2: Overview of Studies Emphasizing Distant User Collaboration

Term	Article	Exemplary Assumptions of Merits
Potential User	Adner, 2002; Arnold et al., 2011; Callahan & Lasry, 2004; Danneels, 2002, 2003, 2004; Day, 1999; De Coster & Butler, 2005; Flores, 1993; Fuchs & Schreier, 2011; Gilbert, 2003; Godes, 2012; Govindarajan et al., 2011; Gruner & Homburg, 2000; Haefliger et al., 2010; Hamel & Prahalad, 1991; Hang et al., 2010; Hauser et al., 2006; Hyatt, 2008; Leonard & Rayport, 1997; Mascitelli, 2000; Mullins & Sutherland, 1998; Prahalad & Ramaswamy, 2000; Reinhardt & Gurtner, 2011; Rothwell et al., 1974; Sawhney et al., 2003; Shaw, 1985; Slater & Naver, 1998; Slater & Mohr, 2006; van den Hende & Schoormans, 2012; Vargo & Lusch, 2004; von Hippel, 1994; Walter et al., 2011	"Market orientation measures do not explicitly probe for a company's exploration of potential customers, but focus on the firm's tight coupling to its current customers." (Danneels, 2003, p. 574) "Increasing a focus on acquiring customers [potential customers] enhances the diversity of customer knowledge development and resource exploration, which relates positively to greater radical innovation performance." (Arnold et al., 2011, p. 244)
New User	Arnold et al., 2011; Bettencourt & Bettencourt, 2011; Danneels, 2002, 2007; Ettlie et al., 1992; Garvin & Levesque, 2006; Henderson, 2006; Jansson, 2011; Johnson et al., 2008; Kumar et al., 2010; Kylaheiko et al., 2011; Lau et al., 2010; Lettice & Parekh, 2010; McGrath et al., 2006; Nijssen et al., 2012; Philipsen et al., 2008; Schmidt & Druehl, 2008; Vorhies et al., 2011; Washburn & Hunsaker, 2011	"[...] radical innovation is always introduced by new firms because new firms focus on new customers and their potential needs, not existing customers [...]. Existing firms that tightly serve or co-develop their products with current key customers may take the risk of being blindsided by a new generation of technology and market niches." (Lau et al., 2010, p. 771)
Emerging User	Day, 1999; Govindarajan & Kopalle, 2004, 2006a, 2006b; Govindarajan et al., 2011; Hoffmann et al., 2010; Slater & Mohr, 2006	"[...] an SBU's emerging-customer-orientation may be an ability that could drive disruptive innovation." (Govindarajan & Kopalle, 2004, p. 5)
Non-User	Christensen, 2006; Christensen & Bower, 1996; Danneels, 2003	"The right lead customers for sustaining innovations are different from those for disruptive innovations. The lead users for new-market innovations may not yet be users." (Christensen, 2006, p. 51)
Future User	Bond & Houston, 2003; Chandy & Tellis, 1998	"The negative findings about market orientation [...] refer to firms that stayed close to their current markets. Our field interviews with radically innovative firms suggest that such firms focus on future customers [...]." (Chandy & Tellis, 1998, p. 479)
Prospective User	Anthony, 2009; Danneels, 2003	"[...] serving current customers is a fundamentally different activity from exploring prospective customers." (Danneels, 2003, p. 574)
Distant User	Jespersen, 2010	"The larger this distance [cognitive distance] is, the more novel information is contained in user inputs." (Jespersen, 2010, p. 472)

Their needs might even be unfulfilled and their contributions might therefore drive change in dominant designs. Chandy and Tellis (1998, p. 479) claim that potential users cause "decision makers in a firm to become keenly aware of market-related developments." It is a deliberate search and scanning of shifts in needs, social norms and values that define the boundaries of a market and its participants. Potential users might be better in reflecting these developments than the current users at the center of the established domain. Lau et al. (2010) arrive at an even more drastic conclusion stating that co-development with current users does not lead to radical changes and that either potential or lead user collaboration is the dominant strategy in early product development. However, this conclusion is derived from the finding that current user co-development does not correlate with product innovation. The merit of collaboration with potential or lead users is only stated as a suggestion requiring further corroboration. However, empirical studies on the micro-level addressing the knowledge intensity as a means to classify users are few (e.g. Kristensson et al., 2008; Magnusson, 2009) and do not cross the boundaries of a target market to reach users at the peripheries and beyond. No systematic inquiry into their unique contribution and how it can be best extracted has been carried out to date. As to the managerial importance of a well-structured exploration towards new market opportunities, a systematic inquiry into the collaboration with potential users presents itself an insightful research objective.

2.2.3 Why Discontinuous Innovation?

Garcia and Calantone (2002) have advanced researchers' understanding of innovations by unifying and consolidating the plethora of definitions and measures used to describe innovation types and degrees that have arisen over the last decades in research on innovation management. The definitions and measures in their meta-analysis have one thing in common: innovativeness has always been modeled as a degree of discontinuity in market related and/or technology related factors. Innovations stimulate new markets to evolve or extend the boundaries of the current ones. They reflect changes in technological capabilities within an industry or domain (Kaplan, 1999; Veryzer, 1998). Discontinuous innovations are classified to result in one or both of these two dimensions of discontinuity, while an innovation to be classified as a radical innovation would require discontinuity in both (Garcia & Calantone, 2002). A new product or service which is used in a different contextual setting than those common to established products in a domain is an example for a discontinuous innovation. It involves changes in the behavior pattern of current users of a domain and additionally a focus on a different type of early users. Organizations involved within the industry therefore need to adapt their market competences towards these emergent segments. The technological requirements immanent to the product or service do not necessarily change. But how can organizations anticipate these changes in usage contexts, user benefits or even shifts in needs and social norms in their domain? Irregularities and anomalies emerge from individuals who are not strongly implicated in current standards and designs. The capacity to learn and prepare for discontinuities resides in ill-defined market structures and at the peripheries of established markets and requires unconventional approaches to be unearthed (Lynn et al., 1996; Reid & Brentani, 2004; Utterback, 1995). Potential users, who might gain importance to a domain on a long-term horizon, can stimulate a crucial and meaningful learning process for organizations trying to explore new markets. This reasoning makes intuitive sense and might explain why researchers in innovation management abundantly praise the merits of collaboration with potential users. It is also, however, apparent that collaboration with potential users is accompanied by many uncertainties. A micro level inquiry is therefore needed, which acknowledges the users' subjectivity in terms of their contextual and experiential contributions in order to fully understand the merits of collaboration with potential users in more detail. For indeed, it is the user's subjectivity that can uncover potential technological applications for discontinuous innovations.

2.2.4 User Classification in Empirical Studies

Only very few studies to date have attempted to validate the foundations behind the benefits drawn from collaboration with potential users (see Table 3). In the course of the literature review, it became increasingly obvious that defining who will be interesting users in the near-by future is as difficult as depicting the next game-changing technologies. The potential users and their needs will always be partly ill-defined and not well-known from the present perspective. Empirical studies have therefore mostly focused on a particular dimension to classify individuals. Kristensson et al. (2004)

showed that ideas generated by ordinary users scored higher on a novelty and creativity level than those generated by professional developers or advanced users. In their study user types were differentiated according to their academic education in programming. According to their typology, the ordinary user is less skilled compared to professional developers or advanced users, but only in a relative sense. The ordinary user lacks the procedural knowledge of how to arrive at a problem solution. The developers and advanced users, by contrast, have a clearer understanding of strategies towards solutions and therefore tend to engage in reproductive and not in productive thinking (Ekvall, 1997). Their expertise encompasses professional training and conscious learning and results in a more structured knowledge pertaining to organization and representation (Chi et al., 1982). However, all these users still take part in the respective market. Chandy and Tellis (1998) state that the orientation to future markets in their case can either refer to the involvement of a different group of people than current users, or explore future needs of current users. The former is not defined in more detail. Bonner and Walker (2004) classify the distance of users by frequency of collaboration (embeddedness), with the assumption that the contribution of something radically new seems to decrease with the frequency of contacts. Hang et al. (2010) refer to a slightly different classification of potential users along the economical pyramid in developing countries. The bottom of the pyramid, being the poorest socio-economic group, is suspended from taking part in a market due to low economical endowment or not meeting circumstance requirements. Organizations willing to accommodate to the particulars of these markets might face vast and untapped opportunities for growth. Others do not use a classification, but leave it to the interviewed managers of business units to decide whether their user involvement is rather directed towards their current or towards potential users with no further explanation of the differentiation. In the following paragraph we will introduce and suggest a scheme for user classification that is in line with the previously given definition of a potential user.

Table 2.3: Empirical Studies Classifying Users

Study	Focus*	Inquiry Subject	User Classification	Result
Chandy & Tellis, 1998	B2C	Willingness to cannibalize and radicalness of innovation	SBU level questionnaire regarding orientation towards future users and needs	Orientation positively influences willingness to cannibalize and thereby radical innovation
Arnold et al., 2011	B2C	Radical innovation performance	Level of customer acquisition/retention orientation in SBU	Firm's that focus on customer acquisition (retention) enhances (worsen) its radical innovation performance
Govindarajan et al., 2011	B2C	Disruptiveness of innovation	SBU level questionnaire regarding orientation towards emergent users	Orientation positively influences the releases of disruptive innovation
Bonner & Walker, 2004	B2B	New product advantage	Relational embeddedness and knowledge heterogeneity	New product advantage tended to be higher in projects that involved less embedded users
Kristensson et al., 2004	B2C	Originality, value and realization of an idea	Academic education in programming	Higher scores on originality and value of ideas from subjects with less programming skills
Kristensson et al., 2008	B2C	Originality of idea	n/a	Limited expertise is not a barrier to useful creative thinking
Lau et al., 2010	B2C	New Product Performance (NPP)	Only considered current users	Current users did not improve NPP; therefore companies should only co-develop with potential or lead users
Kristensson & Magnusson, 2010	B2C	Radical nature of ideas	Users' awareness of technological restrictions	Users who are unaware of any technological restrictions tend to produce more radical service ideas
Hang et al., 2010	B2C	Disruptiveness	Bought or did not buy a unit	Disruption can be caused by addressing the low-end of the economical pyramid in developing countries

*B2B = Business-to-Business; B2C = Business-to-Consumer

2.3 Proposal for User Classification

It seems plausible that there are a variety of dimensions which might constitute a person's distance with regard to a particular domain. It is conceptually hard to operationalize and foresee within a population which people will become important to a domain and its organizations. We propose to focus, therefore, on two dimensions, an experience and a market dimension. The former relates to the past interaction and experience record of an individual and the latter to the general outlook of an individual with respect to a particular domain, regardless of whether the person is or is not using a product or service of the domain.

Experience dimension. In the constructivist tradition, knowledge is subjective (Scott et al., 1979). It constitutes ideas a person holds about the self, the world or objects and is strongly influenced by a person's individual experience record (Hunt, 2003). Validated constructs in literature that relate to users' knowledge are e.g. *use experience* and *user knowledge* (Alba & Hutchinson, 1987; Park et al., 1994). These are by no means independent of each other, but they still relate to different aspects of knowledge. Use experience constitutes knowledge through direct interaction between the user and a product or service. Cognitive structure originates from this process. The interaction creates mental models relating to the domain and generates various logics of interaction between performance and components of a product or service (Mitchell & Dacin, 1996). These cognitive structures provide thereafter the basis for further involvement with the domain, whether pertaining to a purchase decision or in a creative process (Alba & Hutchinson, 1987; Ward, 1995). In contrast, user knowledge presents the body of knowledge on a broader scope. It encompasses sources other than usage and might be classified as being more theoretical expertise (Brucks, 1985). Even potential users can gain user knowledge to a certain degree, for instance by being exposed to commercials, reading about a domain in magazines or through interaction with peers.

Market dimension. Market segmentations of companies mostly build on demographic variables, such as age and income. These segmentations do not, however, reflect the problems people attempt to solve by using a product and are thereby not very useful in attempting to uncover upcoming market participants (Christensen et al., 2007). Furthermore, the ownership of a product or the access to a service is insufficient to represent the affiliation of an individual to a market. It is also more related to the construct of use experience (Alba & Hutchinson, 1987; Park et al., 1994). A dimension that represents the affiliation or connectedness of a person with a domain seems more appropriate, since it allows for finer grained distinctions including people who might connect with a domain even though they are not current users. Remote and upcoming needs, entrance barriers for users or new application contexts indicate relevant market development areas. Ideas that have disruptive potential nurture, for instance, in remote and niche market segments (Christensen, 1997; Henderson, 2006). The needs of people in these niches are often directed towards different performance parameters of particular product classes than those valued by the current market (Danneels, 2004). Shifts of interest from the current market towards the niche irreversibly change the market structure and make the study of the particulars of the edges of a market insightful or even necessary to survive (Henderson, 2006). The *involvement* construct (Zaichkowsky, 1985) is appropriate here in that it can be drawn upon to identify those people who - although they may not actually be using a product or service - hold a positive tenor towards the domain. It takes into account the person's inherent interests, values, and needs.

To sum up, the items of the use experience construct define whether someone is actually a current user or non-user and how frequently an individual interacts with a product or service. The items of the user knowledge construct allow for a fine-grained differentiation with regard to the characteristics of domain products, trends within the market and their knowledge in relation to others. The items of the involvement construct then inquire whether an individual is personally interested in a domain, regardless of whether he or she is a current or non-user. Taken together, these constructs can distinguish current from non-users as well as split non-users into potential users and the vast part of the population.

2.3.1 Mapping the Dimensions

We assume the experience and market dimensions to influence each other, just as needs influence an individual's perception by affecting the availability of interpretive categories (Scott et al., 1979). Obviously, high involvement with a certain object might drive a person to get involved with this object, which leads to personal experience. On the other hand, gaining experience with an object through interaction can create involvement e.g. by awakening latent needs. When mapping the dimensions and defining both by a distance measure increasing from its point of origin, the result is the segmentation as depicted in Figure 2.1. Three rough segments emerge along the dimensions: the current user, the potential user and the vast population of non-users. The dashed line emblematizes a well-balanced relation between these two dimensions. The proportion of the two distances to each other might be useful to refine each segment in more detail. The segment labeled vast population is a necessity to account for the part of the population that does not want to take part in a market. Quite often, non-users are just that (Anthony et al., 2008). Current users are those who use the products and services of a particular domain on a regular basis. They have a need that is fulfilled by a particular class of products and matches that which the respective organization is trying to address. Experience is gained by constant usage and constructs domain relevant knowledge pertaining to performance attributes, (physical) components and how (physical) components and attributes affect performance attributes (Mitchell & Dacin, 1996). Furthermore, current users value certain proportions between the attributes of a product and can distinguish between objects by indicating whether an object has more or less of an attribute (Alba & Hutchinson, 1987; Mitchell & Dacin, 1996). The focus on the most valued performance attributes of dominant designs might obstruct the perspectives for discontinuous changes (Christensen, 1997). Current users use a product or service for a particular purpose and might not be willing to dispense on the attribute-performance proportion of crucial components. Behavior of this kind resonates with the consistency theories explained earlier on. The emergence of new dimensions or emphasis on a rather unvalued product dimension, however, can be an important stimulus to discover new market opportunities and application contexts (Christensen, 1997, 2006; Govindarajan & Kopalle, 2006; Henderson, 2006).

Nintendo's Wii, for instance, revolutionized the market of game consoles. Unlike the established consoles available at the time of its market rollout, it did not focus on meticulous detail in the graphics and control of its characters. Its focus was rather on simple graphics, decreasing access barriers by introducing simplicity in interaction patterns[2], and making game consoles socially engaging. It is unlikely that an experienced and adept gamer would have suggested a downgrade in the level of graphic detail simply in order to make the game's purpose more intuitively understandable and accessible. This points to the importance of exploring the surrounding segment.

---

[2] In fact the interaction pattern with the Wii control is an identical representation of established movements in the "analogue" version of its games, e.g. swinging your arm to play tennis.

Figure 2.1: Mapping the Experience and Market Distance of Users

The people who belong to the segment labeled potential users are on one or both dimensions significantly further on the scale than the current users. Around the dashed-line, the proportion of the dimensions is balanced, suggesting less involvement goes hand in hand with less experience and vice-versa. But how are we to interpret the extremes to the upper-left or lower-right of that line, as depicted in the hatched quadrates in Figure 2.2.? In the upper left quadrate the market distance is proportionally lower to the experience distance. Here the potential user is associated with a high involvement - e.g. in the form of a distinct need - with regards to the product class under consideration, but he or she lacks personal experience. These potential users are the ones depicted in the definition of Hauser et al. (2006). One explanation for the imbalance between the dimensions might be a constraint the user faces, as represented by one or more usage barrier(s). The barriers these users might face, however, are not congruent with the ones discussed in the literature dealing with resistance to innovations (Garcia et al., 2007; Ram & Seth, 1989). Dominant designs might have reached the level of being a commodity altogether and be all the more on the verge to a discontinuous change. Potential risk or traditional barriers for instance might have been torn down a while ago. In this realm, barriers relating to the endowment of personal characteristics of the user or the dominant characteristics of the particular product class might be a more likely cause for someone's lack of experience. As previously stated, the four most common barriers are lack of skills, means, time or access (Christensen, 2006; Govindarajan & Kopalle, 2006b). In particular cases, the latter is strongly related to the context of common usage. Some people might not use a certain product since it is not available in contexts where they would like to use it. An example would be people's increasingly strong need to pass their time while waiting for a bus, before smart phones enabled people to pass their time on the go as they would do at home (e.g. mobile television). These barriers are rigid as they cannot easily be removed with marketing efforts. They demand a strong alteration of the product on offer, in some cases even in the underlying business model (Christensen, 2006). Collaborating with potential users early on might shed light on new and unexplored usage contexts for an existing technology. Alternatively, organizations might stumble upon different ways to approach a need, since potential users might have alternative ways to meet their need in the end. In the lower right quadrate, experience distance is proportionally lower to the market distance. Here the potential user has pristine experience with regards to the product class under consideration, but lacks a distinct interest or need. One explanation for this imbalance might be unorthodox usage (Hentschel, 2009). It is assumed that products have one central function, but in fact users do not always use the product for that purpose or in the way it was intended to be used by the organization (Bercun, 2007). Often product attributes and functions are used to meet a slightly different need or in extreme cases are applied in a whole other domain. Well-known ideas have emerged when product functions have been applied to other domains. Thus, a vine press was used for book printing, white paint was transformed into a means of erasing typing errors when mechanically typewriting (Hentschel, 2009). This originally unorthodox usages were triggered by the absence of an appropriate product or by inflicted barriers such as monetary endowment or restrained access. Indeed generally, organizations tend to have but a limited view on how their products end up being used (Hentschel, 2009), but changes of the product itself or the application context might help loosen what is often referred to as inflexibility due to established mental models (Allen & Marquis, 1964).

Figure 2.2: Two Extremes in the Potential User Segment

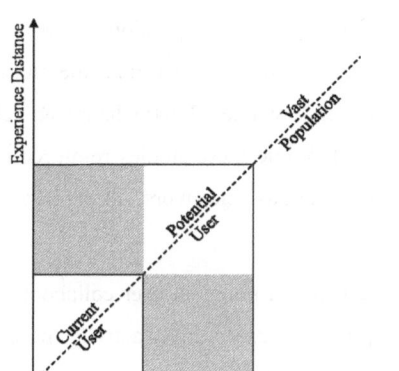

A user who addresses a different need with a product than intended by the organization is more distant to the domain than a current user. From an organizational perspective, potential users can point towards new application scenarios for a product or service as well as new domains to get involved in.

In both hatched quadrates in Figure 2.2 the user might be confronted with barriers. The ones in the upper left quadrate prevent people from using a product or service of a particular domain. The ones in the lower right quadrate exist in other domains and might force people to use a product or service in an unorthodox way. Both directions seem to allow for insightful and imaginative paths to pursue in organizational exploration. Throughout the literature, it remains unclear, however, which references frames potential users base their contributions on and how their contributions affect a collaborative process. A more detailed and comprehensive research approach would be required to decode the gains that can be drawn from collaboration with distant informants. The next paragraph is an attempt to summarize these research gaps.

2.4 Identification of Research Gaps

A careful selection of the right external partners in the development of highly innovative products and services is crucial (Fischer & Reuber, 2004). In the context of collaboration with users, the issue of selection has not yet been sufficiently addressed (Lettl, 2007; Nijssen et al., 2012). Current empirical investigations into the virtue of collaboration with potential users tackle the question on an organizational level, using structured interview data from business units and the success measures of final innovation in the market (Bonner & Walker, 2004; Chandy & Tellis, 1998; Govindarajan et al., 2011; Hang et al., 2010). While these studies reinforce the idea of collaboration with potential users in discontinuous endeavors, three important gaps still call for clarification.

Firstly, a validation of the merit of potential user collaboration on the organizational level might not be appropriate in explorative activity based on collaboration and collective action. At this point, users' inputs are inseparably bound up with the organizational efforts and the efficiency of co-creation is therefore hard to grasp. The benefits of user collaboration are generally considered to arise from probing into their subconscious knowledge, life goals and motivations (Danneels, 2003; Day, 1999; Mascitelli, 2000; Visser et al., 2007; von Hippel, 1994). This information usually needs to be extracted upstream in the innovation process by a few key individuals of the organization (Reid & Brentani, 2004). The value of potential users would need to be evaluated at its point of origin. When talking about explorative collaboration, the outcome of the collaboration needs to be analyzed.

Secondly, the existing empirical studies do not examine the constituents of the unique outputs of potential users in terms of innovativeness and discontinuity (Danneels & Kleinschmidt, 2001). Differentiation into the widely acknowledged dimensions of innovativeness - innovations causing market and/or technological discontinuity - is a fruitful analysis (Garcia & Calantone, 2002). It indicates how the merits of potential user collaboration relate to the competences of an organization.

Thirdly, and most importantly, the results of the empirical studies do not explain the unique contributions brought about by collaborations with potential users as opposed to current users and how these are be made explicit in order to be understood by the right people inside an organization. In fact, they dismiss the collaborative perspective all together, though it has become the core of innovative activities nowadays (Dunbar, 1997;

Leonard & Sensiper, 1998; Reid & Bentrani, 2004). Conceptually it is not apparent how potential users influence the innovative activity or how they are affected by the collective action themselves. Even in the literature on the front-end of innovation, the processes and contents of idea generation activities in collaboration and how these relate to innovations have received little attention (Kristensson & Magnusson, 2010). A study of the micro processes when collaborating with potential users is therefore urgently needed. It should pay deliberate attention to the unique context of the collaboration, the collaborative processes and the inherent communication to deliver a rich and comprehensive picture. Only then can guidelines for organizational exploration in collaboration with the edges of their markets be derived and made available.

In sum, the idea of absorbing more discontinuous stimuli for innovation by means of collaboration with potential users requires corroboration at its point of origin, explanation in terms of content and process, and elaboration towards practical guidelines for a more clearly steered exploration (Teddlie et al., 2008). Statistical results are needed (validation), as well as an in-depth analysis of individual contributions and a thorough investigation into the contexts in which they occur (Clark et al., 2008).

2.5 Discussion and Conclusion

New usage contexts and upcoming needs might upset the existing order of established practice. To hit upon these impulses, it would be fruitful for an organization to explore the borders of markets (Chandy & Tellis, 1998; Danneels, 2003; Christensen, 1997). In this paper, we have derived a comprehensive definition of the potential user from research on innovation management. We have explained why the virtue that can be expected to arise in collaboration with potential users corresponds well to the definition of discontinuous innovation drawn from merging theories and insights from research in the sociocognitive psychology of creativity, degrees of innovativeness and user involvement. It remains crucial that the expected outcome of innovative activities be made explicit in order for a common understanding of the terms used to be achieved and for different studies to be made comparable (Garcia & Calantone, 2002). Applying the constructs of use experience, user knowledge and involvement to the context of user collaboration in exploration, the definition of a potential user has been operationalized. A mapping of these dimensions has explored propositions to explain why collaboration with potential users might generate useful insight for change and why it might cause discontinuity on the market but not on the technological competences of an organization. The conceptualization of collaboration with potential users will be the basis for subsequent empirical studies.

References

Adner, R. 2002. When are Technologies Disruptive? A Demand-Based View of the Emergence of Competition. *Strategic Management Journal*, 23(8): 667-688.

Alba, J.W., & Hutchinson, W.J. 1987. Dimensions of Consumer Expertise. *Journal of Consumer Research,* 13(4): 411–454.

Allen, T. J., & Marquis, D. G. 1964. Positive and Negative Biasing Sets: The Effects of Prior Experience on Research Performance. *IEEE Transactions on Engineering Management*, 11(4): 158–61.

Anthony, S. D. 2009. Major League Innovation. *Harvard Business Review*, 87(10): 51-54.

Anthony, S., Johnson, M., Sinfield, J. & Altman, E. 2008. *The Innovator's Guide to Growth.* Boston, Harvard Business Press.

Arnold, T. J., Fang, E., & Palmatier, R. W. 2011. The Effects of Customer Acquisition and Retention Orientations on a Firm's Radical and Incremental Innovation Performance. *Journal of the Academy of Marketing Science*, 39(2): 234-251.

Bercun, S. 2007. *The Myths of Innovation.* O'Reilly, Sebastopol, CA.

Bettencourt, L. A., & Bettencourt, S. L. 2011. Innovating on the Cheap. *Harvard Business Review*, 89(6): 88-94.

Bharadwaj, N., Nevin, J. R., & Wallman, J. P. 2012. Explicating Hearing the Voice of the Customer as a Manifestation of Customer Focus and Assessing its Consequences. *Journal of Product Innovation Management*, forthcoming.

Bogers, M., Afuah, A., & Bastian, B. 2010. Users as Innovators: A Review, Critique, and Future Research Directions. *Journal of Management,* 36: 857–875.

Bond, E., & Houston, M. 2003. Barriers to Matching New Technologies and Market Opportunities in Established Firms. *Journal of Product Innovation Management*, 20(2): 120-135.

Bonner, J. M., & Walker Jr., O. C. 2004. Selecting Influential Business-to-Business Customers in New Product Development: Relational Embeddedness and Knowledge Heterogeneity Considerations. *Journal of Product Innovation Management*, 21: 155-169.

Brucks, M. 1985. The Effects of Product Class Knowledge on Information Search Behavior. *Journal of Consumer Research,* 12(1): 1–16.

Callahan, J., & Lasry, E. 2004. The Importance of Customer Input in the Development of Very New Products. *R&D Management,* 34(2):107–120.

Chandy, R. K., & Tellis, G. J. 1998. Organizing for Radical Product Innovation: The Overlooked Role of Willingness to Cannibalize. *Journal of Marketing Research,* 35(4): 474–487.

Chi, M., Glaser, R., & Rees, E. 1982. Expertise in Problem-solving. In R. S. Sternberg (Ed.), *Advances in the Psychology of Human Intelligence,* vol. 1: 1-75. Hillsdale, NJ Erlbaum.

Christensen, C. 1997. *The Innovator's Dilemma.* Boston: Harvard Business School Press.

Christensen, C. 2006. The Ongoing Process of Building a Theory of Disruption. *Journal of Product Innovation Management,* 23: 39-55.

Christensen, C., Anthony, S. D., Berstell, G., & Nitterhouse, D. 2007. Finding the Right Job For Your Product. *Sloan Management Review,* 48: 37-47.

Christensen C. M., & Bower, J. L. 1996. Customer Power, Strategic Investment, and the Failure of Leading Firms. *Strategic Management Journal,* 17(3): 197–218.

Clark, V. L. P., Creswell, J. W., Green, D. O., & Shope, R. J. 2008. Mixing Quantitative and Qualitative Approaches: An Introduction to Emergent Mixed Methods Research. In S. N. Hesse-Biber & P. Leavy (Eds.), *Handbook of Emergent Methods:* 363-388. New York: Guilford.

Danneels, E. 2002. The Dynamics of Product Innovation and Firm Competences. *Strategic Management Journal,* 23: 1095-1121.

Danneels, E. 2003. Tight-Loose Coupling with Customers: The Enactment of Customer Orientation. *Strategic Management Journal,* 24: 559-576.

Danneels, E. 2004. Disruptive Technology Reconsidered: A Critique and Research Agenda. *Journal of Product Innovation Management,* 21: 246–258.

Danneels, E. 2007. The Process of Technological Competence Leveraging. *Strategic Management Journal,* 28(5): 511-533.

Danneels, E., & Kleinschmidt, E. J. 2001. Product Innovativeness from the Firm's Perspective: Its Dimensions and their Relation with Project Selection and Performance. *Journal of Product Innovation Management,* 18(6): 357-373.

Day, G. S. 1999. Misconceptions about Market Orientation. *Journal of Market Focused Management,* 4: 5-16.

De Coster, R., & Butler, C. 2005. Assessment of Proposals for New Technology Ventures in the UK: Characteristics of University Spin-off Companies. *Technovation,* 25(5): 535-543.

Dunbar, K. 1997. How Scientists Think: On-Line Creativity and Conceptual Change in Science. In T. B. Ward, S. M. Smith & J. Vaid (Eds.), *Conceptual structures and processes: Emergence, Discovery, and Change:* 461-493: Washington D.C: American Psychological Association Press.

Duncker, K. 1945. On Problem-solving. *Psychological Monographs,* 58(5): whole no. 270.

Ekvall, G. 1997. Organizational Conditions and Levels of Creativity. *Creativity and Innovation Management,* 6: 195-205.

Ettlie, J., & Reza, E. 1992. Organizational Integration and Process Innovation. *Academy of Management Journal,* 35(4): 795-827.

Festinger, L. 1957. *A Theory of Cognitive Dissonance.* Stanford, CA: Stanford University Press.

Fischer, E., & Reuber, A. R. 2004. Contextual Antecedents and Consequences of Relationships between Young Firms and Distinct Types of Dominant Exchange Partners. *Journal of Business Venturing,* 19(5): 681–706.

Flores, F. 1993. Innovation by Listening Carefully to Customers. *Long Range Planning,* 26(3): 95-102.

Foxall, G. R. 1989. User-initiated Product Innovations. *Industrial Marketing Management,* 18(2): 95–104.

Fuchs, C., & Schreier, M. 2011. Customer Empowerment in New Product Development. *Journal of Product Innovation Management,* 28(1): 17-32.

Garcia, R., Bardhi, F., & Friedrich, C. 2007. Overcoming Consumer Resistance to Innovation. *Sloan Management Review,* 48: 82-88.

Garcia, R., & Calantone, R. 2002. A Critical Look at Technological Innovation Typology and Innovativeness Terminology: A Literature Review. *Journal of Product Innovation Management,* 19(2): 110–132.

Garvin, D. A., & Levesque, L. C. 2006. Meeting the Challenge of Corporate Entrepreneurship. *Harvard Business Review,* 84(10): 102-112.

Gilbert, C. 2003. The Disruption Opportunity. *Sloan Management Review,* 44(4): 27-32.

Govindarajan, V., & Kopalle, P. K. 2004. Can Incumbents Introduce Radical and Disruptive Innovations? Working Paper, Marketing Science Institute, Cambridge, MA.

Govindarajan, V., & Kopalle, P. K. 2006a. Disruptiveness of Innovations: Measurement and an Assessment of Reliability and Validity. *Strategic Management Journal,* 27: 189-199.

Govindarajan, V., & Kopalle, P. K. 2006b. The Usefulness of Measuring Disruptiveness of Innovations Ex Post in Making Ex Ante Predictions. *Journal of Product Innovation Management,* 23: 12-18.

Govindarajan, V., Kopalle, P. K., & Danneels, E. 2011. The Effects of Mainstream and Emerging Customer Orientations on Radical and Disruptive Innovations. *Journal of Product Innovation Management,* 28: 121-132.

Greer, C. R., & Lei, D. 2012. Collaborative Innovation with Customers: A Review of The Literature and Suggestions for Future Research. *International Journal of Management Reviews,* 14(1): 63–84.

Gruner, K., & Homburg, C. 2000. Does Customer Interaction Enhance New Product Success? *Journal of Business Research,* 49(1): 1-14.

Haefliger, S., Jaeger, P., & von Krogh, G. 2010. Under the Radar: Industry Entry By User Entrepreneurs. *Research Policy,* 39(9): 1198-1213.

Hamel, G., & Prahalad, C. K. 1991. Corporate Imagination and Expeditionary Marketing. *Harvard Business Review,* 69(4): 81–92.

Hang, C., Chen, J., & Subramian A. M. 2010. Developing Disruptive Products for Emerging Economies: Lessons from Asian cases. *Research Technology Management:* 21-26.

Hauser, J. R., Tellis, G. J., & Griffin, A. 2006. Research on Innovation: A Review and Agenda for Marketing Science. *Marketing Science*, 25(6): 687-717.

Henderson, R. 2006. The Innovator's Dilemma as a Problem of Organizational Competence. *Journal of Product Innovation Management*, 23: 5-11.

Hentschel, C. 2009. Attribute-Domain Matrix: A Reverse Engineering Method for Innovation. *Creativity and Innovation Management*, 18: 81-89.

Hoffman, D. L., Kopalle, P. K., & Novak, T. P. 2010. The Right Consumers for Better Concepts: Identifying and Using Consumers High in Emergent Nature to Further Develop New Product Concepts. *Journal of Marketing Research*, 47: 854-865.

Hunt, D. P., 2003. The Concept of Knowledge and how to Measure it. *Journal of Intellectual Capital*, 4: 100-113.

Hyatt, J. 2008. The Incrementalist (or, What's the Small Idea?). *Sloan Management Review*, 49(4): 15-20.

Jansson, J. 2011. Emerging (Internet) Industry and Agglomeration: Internet Entrepreneurs Coping with Uncertainty. *Entrepreneurship and Regional Development*, 23(7-8): 499-521.

Jespersen, K. R. 2010. User-Involvement and Open Innovation: The Case of Decision-Maker Openness. *International Journal of Innovation Management*, 14: 471-489.

Johnson, M. W., Christensen, C. M., & Kagermann, H. 2008. Reinventing your business model. *Harvard Business Review*, 86(12): 50-59.

Kanter, R. M. 2007. Der sichere Pfad zu Innovationen. *Harvard Business Manager*, 2, 44–63.

Kaplan, S. M. 1999. Discontinuous Innovation and the Growth Paradox. *Strategy & Leadership*, 27: 16 – 21.

Kristensson, P., Gustafsson, A., & Archer, T. 2004. Harnessing the Creative Potential among Users. *Journal of Product Innovation Management*, 21: 4-14.

Kristensson, P., & Magnusson, P. R. 2010. Tuning Users' Innovativeness During Ideation. *Creativity and Innovation Management*, 19: 147-159.

Kristensson, P., Matthing, J., & Johansson, N. 2008. Key Strategies for the Successful Involvement of Customers in the Co-Creation of New Technology-Based Services. *International Journal of Service Industry Management*, 19(3-4): 474-491.

Kumar, V., Aksoy, L., Donkers, B., Venkatesan, R., Wiesel, T., & Tillmanns, S. 2010. Undervalued or Overvalued Customers: Capturing Total Customer Engagement Value. *Journal of Service Research*, 13(3): 297-310.

Kylaheiko, K., Jantunen, A., Puumalainen, K., Saarenketo, S., & Tuppura, A. 2011. Innovation and Internationalization as Growth Strategies: The Role of Technological Capabilities and Appropriability. *International Business Review*, 20(5): 508-520.

Lau, A. K. W., Tang, E., & Yam, R. C. M. 2010. Effects of Supplier and Customer Integration on Product Innovation and Performance: Empirical Evidence in Hong Kong Manufacturers. *Journal of Product Innovation Management*, 27(5): 761-777.

Lettice, F., & Parekh, M. 2010. The Social Innovation Process: Themes, Challenges and Implications for Practice. *International Journal of Technology Management*, 51(1): 139-158.

Leonard, D., & Rayport, J. F. 1997. Spark Innovation through Empathic Design. *Harvard Business Review*, 75: 102-113.

Leonard, D., & Sensiper, S. 1998. The Role of Tacit Knowledge in Group Innovation. *California Management Review*, 40: 112–132.

Lettl, C. 2007. User Involvement Competence for Radical Innovation. *Journal of Engineering and Technology Management*, 24: 53-75.

Lettl, C., Herstatt, C., & Gemünden, H. G. 2006. Users' Contributions to Radical Innovation: Evidence from four Cases in the Field of Medical Equipment Technology. *R&D Management*, 36: 251-272.

Lüthje, C. 2004. Characteristics of innovating users in a consumer goods field. *Technovation*, 24(9): 1-28.

Lynn, G. S., Morone, J. G., & Paulson, A. S. 1996. Marketing and Discontinuous Innovation: the Probe and Learn Process. *California Management Review*, 38(3): 8-37.

Magnusson, P. R. 2009. Exploring the Contributions of Involving Ordinary Users in Ideation of Technology-Based Services. *Journal of Product Innovation Management*, 26(5): 578–593.

Mascitelli, R. 2000. From Experience: Harnessing Tacit Knowledge to Achieve Breakthrough Innovation. *Journal of Product and Innovation Management*, 17: 179-193.

McGrath, R. G., Keil, T., & Tukiainen, T. 2006. Extracting Value from Corporate Venturing. *Sloan Management Review*, 48(1): 50-56.

Mitchell, A. A., & Dacin, P. A. 1996. The Assessment of Alternative Measures of Consumer Expertise. *Journal of Consumer Research*, 23(3): 219–39.

Montuori, A., & Purser, R. E. 1999 (Ed.): *Social Creativity*, vol. 1: 1-45 Cresskill.

Mullins, J., & Sutherland, D. 1998. New Product Development in Rapidly Changing Markets: An Exploratory Study. *Journal of Product Innovation Management*, 15(3): 224-236.

Nijssen, E. J., Hillebrand, B, de Jong, J. P. J., & Kemp, R. G. M. 2012. Strategic Value Assessment and Explorative Learning Opportunities with Customers. *Journal of Product Innovation Management*, 29(S1): 91-102.

Park, W. C., Mothersbaugh, D. L., & Feick, L. 1994. Consumer Knowledge Assessment. *Journal of Consumer Research*, 21(1): 71–82.

Philipsen, K., Damgaard, T., & Johnsen, R. E. 2008. Suppliers' Opportunity Enactment through the Development of Valuable Capabilities. *Journal of Business & Industrial Marketing*, 23(1): 23-34.

Prahalad, C. K., & Ramaswamy, V. 2000. Co-opting Customer Competence. *Harvard Business Review*, 78(1): 82.

Ram, S., & Sheth, J. N. 1989. Consumer Resistance to Innovations: The Marketing Problem and Its Solutions. Journal of Consumer Marketing 6, no. 2, pp. 5-14.

Reid, S. E., & Brentani, U. 2004. The Fuzzy Front End of New Product Development for Discontinuous Innovations: A Theoretical Model. *Journal of Product Innovation Management,* 21: 170-184.

Reinhardt, R., & Gurtner, S. 2011. Enabling Disruptive Innovations through the Use of Customer Analysis Methods. *Review of Managerial Science*, 5(4): 291-307.

Rosenberg, M. 1956. Cognitive Structure and Attitudinal Affect. *Journal of Abnormal and Social Psychology,* 53: 367-72.

Rothwell, R., Freeman, C., Horlsey, A., Jervis, V. T. P., Robertson, A. B., & Townsend, J. 1974,. Sappho Updated: Project Sappho Phase II. *Research Policy,* 3: 204–225.

Rubenson, D. L., & Runco, M. A. 1995. The Psychoeconomic View of Creative Work in Groups and Organizations. *Creativity and Innovation Management,* 4: 232-241.

Sawhney, M., Prandelli, E., & Verona, G. 2003. The Power of Innomediation. *Mit Sloan Management Review*, 44(2): 77-82.

Sawyer, R. K. 2005. *Social Emergence.* Cambridge University Press, Cambridge.

Schmidt, G. M., & Druehl, C. T. 2008. When is a Disruptive Innovation Disruptive? *Journal of Product Innovation Management,* 25(4): 347–369.

Scott, W. A., Osgood, D. W., Peterson, C. 1979. *Cognitive Structure: Theory and Measurement of Individual Differences.* V.H. Winston & Sons, Washington.

Shaw, B. S. 1985. The Role of the Interaction between the User and the Manufacturer in Medical Equipment Innovation. *R&D Management,* 15: 283-292.

Slater, S. F., & Mohr, J. J. 2006. Successful Development and Commercialization of Technological Innovation: Insights Based on Strategy Type. *Journal of Product Innovation Management,* 23: 26-33.

Slater, S. F., & Narver, J. C. 1998. Customer-led and Market Oriented: Let's not Confuse the Two. *Strategic Management Journal,* 19(10): 1001–1006.

Sonnenburg, S. 2004. Creativity in Communication: A Theoretical Framework for Collaborative Product Creation. *Creativity and Innovation Management,* 13: 254-262.

Teddlie, C., Tashakkori, A., & Johnson, B. 2008. Emergent Techniques in the Gathering and Analysis of Mixed Method Data. In S. N. Hesse-Biber & P. Leavy (Eds.), *Handbook of Emergent Methods:* 389-413. New York: Guilford.

Utterback, J. M. 1996. *Mastering the Dynamics of Innovation.* Harvard Business School Press, Boston, Mass.

van den Hende, E. A., & Schoormans, J. P. L. 2012. The Story is as Good as the Real Thing: Early Customer Input on Product Applications of Radically New Technologies. *Journal of Product Innovation Management,* 29(4): 655-666.

Vargo, S. L., & Lusch, R. F. 2004. Evolving to a New Dominant Logic for Marketing. *Journal of Marketing,* 68(1): 1-17.

Veryzer, R. W. 1998. Key Factors Affecting Customer Evaluation of Discontinuous New Products. *Journal of Product Innovation Management,* 15(2):136–50.

Visser, F. S., van der Lugt, R., & Stappers, P. J. 2007. Sharing User Experiences in the Product Innovation Process: Participatory Design Needs Participatory Communication. *Creativity and Innovation Management,* 16: 35–45.

von Hippel, E. 1986. Lead Users: A Source of Novel Product Concepts. *Management Science,* 32(7): 791–805.

von Hippel, E. 1994. Sticky Information and the Locus of Problem-solving: Implications for Innovation. *Management Science,* 40: 429-439.

Vorhies, D. W., Orr, L. M., & Bush, V. D. 2011. Improving Customer-Focused Marketing Capabilities and Firm Financial Performance via Marketing Exploration and Exploitation. *Journal of the Academy of Marketing Science,* 39(5): 736-756.

Walter, A., Parboteeah, K. P., Riesenhuber, F., & Hoegl, M. 2011. Championship Behaviors and Innovations Success: An Empirical Investigation of University Spin-Offs. *Journal of Product Innovation Management,* 28(4): 586-598.

Ward, T. B. 1995. What's Old About New Ideas? In S. M. Smith, T. B. Ward & R. A. Finke (Eds.), *The creative cognition approach:* 157-178. Cambridge, MA: MIT Press.

Washburn, N. T., & Hunsaker, B. T. 2011. Finding Great Ideas in Emerging Markets. *Harvard Business Review,* 89(9): 115-120.

Weisberg, R. W. 1999. Creativity and Knowledge: A Challenge to Theories. In R. J. Sternberg (Ed.), *Handbook of Creativity:* 226-250. New York: Cambridge University Press.

Zaichkowsky, J. L. 1985. Measuring the Involvement Construct. *The Journal of Consumer Research,* 12: 341-352.

3 Innovative Ideas through Collaboration with Potential Users

Martin Hewing and Katharina Hölzle

Abstract Organizations increasingly use environmental stimuli and ideas from users within participatory innovation processes in order to tap new sources of knowledge. The research presented in this article focuses on users who shape the distant edges of markets and currently are not using products and services from a domain – so called potential users. Those users at the peripheries are perceived to contribute more novel information, by which they better reflect shifts in needs and behavior than current users in the core market. Their contributions in collaborative and creative problem-solving processes and how they generate ideas for discontinuous innovations are of particular interest. With an experimental design, we compare ideas from potential and current users and analyze the effects of cognitive distance in collaboration and the utilization of explicit and tacit knowledge. We find potential users to generate more original ideas, particularly when they collaborate with someone experienced within the domain. Their ideas are most obviously characterized by an increased level of surprise and unusualness compared to dominant designs, which is rooted in contexts and does not require technological leaps. Collaboration with potential users can therefore result in new ways to leverage technological competences. Furthermore, the cross-fertilization arising from cognitive distance between a potential and a current user is asymmetric due to differences in the nature of their utilized knowledge and personal objectives. This paper discusses implications for innovation research and the management of early innovation processes.

3.1 Introduction

Collaboration with users is one of the most often used forms of collaboration with external exchange partners in organizational innovation activities (Nijssen et al., 2012). In the creative economy, sourcing environmental stimuli and ideas from users avoids mono-causal reasoning about needs, problems and solutions early on. Most often, the notion of user involvement is linked to existing market majorities. However, especially in the early idea stage, organizations could prevent incrementalism by focusing on those users they are currently not serving (Christensen, 2006; Danneels, 2003; Lau et al., 2010). Those users at the peripheries are perceived to contribute more novel information, by which they better reflect shifts in needs and behavior than current users (Chandy & Tellis, 1998). Such environmental stimuli are a primary source of ideas for discontinuous innovation and can provide organizations with a competitive edge in growing markets (Crossan et al., 1999; Govindarajan et al., 2011; Reid & Brentani, 2004; Visser et al., 2007). The literature on innovation management makes use of a variety of terms for users who are not within the current target market, such as non-users (e.g. Christensen, 2006; Christensen & Bower, 1996; Danneels, 2003), potential users (e.g. Bonner & Walker, 2004; Fuchs & Schreier, 2011; Shaw, 1985), new users (e.g. Arnold et al., 2011; Danneels, 2002; Henderson, 2006), emerging users (e.g. Day, 1999; Govindarajan et al., 2011; Hoffman et al., 2010; Slater & Mohr, 2006), prospective users (Anthony, 2009; Danneels, 2003) or future users (Bond & Houston, 2003; Chandy & Tellis, 1998). Most of these terms are used interchangeably and definitions are few. In a similar manner to Hauser et al. (2006), we define these distant characters - using the term potential users - as *people, who do not yet use a product or service, but who hold a positive tenor and interest towards the domain.*

Empirical studies concerned with potential user involvement focused on the organizational level and underlined the effects on the innovation performance of business units (Arnold et al., 2011; Bonner & Walker, 2004; Chandy & Tellis, 1998; Govindarajan et al., 2011). Though researchers seem to agree that the merits of collaboration with potential users arise by harnessing their subconscious knowledge and experience (e.g. Kristensson et al., 2008; von Hippel, 1986), approaches focusing on the individual and the micro processes are absent. Yet highly tacit knowledge exchanges might best be researched further upstream in the innovation process on an individual level, preferably even before a project has been formalized on an organizational level (Reid & Bentrani, 2004). So far, collaboration with potential users as individuals remains a vague concept, resulting in a lack of guidelines for how to involve and learn from them. Accordingly, fallbacks into either a sole focus on current users or into traditional market research methods have become a popular choice in organizational practice (Christensen, 1997; Leonard-Barton, 1992). This directly relates to the struggle of organizations to turn away from dominant designs in explorative activities. Often, well established organizations lose sight of their initial goal to explore new markets in favour of more predictable changes or rely on knowledge and processes that have worked well in the past (Audia & Goncalo, 2007; Birkinshaw et al., 2007). Particularly in high-technology industries, in which domain-related expertise has a short half-life, learning from the edges of one's own domain has become crucial. Collaborating with people with distant knowledge sources, however, also leads to the challenge of mutual understanding (Nooteboom, 2000, 2009). This difficulty makes a thorough understanding of processes and ways how to make users' vast potential accessible all the more important. In order to successfully manage creativity in user collaboration, it is necessary to understand the conditions under which users, both potential and current ones, will tend to generate ideas that are discontinuous as opposed to continuous (Audia & Goncalo, 2007). The main research questions are: (1) *Are the ideas contributed by potential users more discontinuous to current market characteristics than those of current users?* (2) *Do the ideas brought in by potential users constitute a greater challenge to technological competences within a domain than those of current users?* (3) *Does the cognitive distance between collaborating users influence their ideas?*

Within an experimental design, the present research analyzes and compares the direct output from potential and current users who engaged in collaborative and creative problem-solving tasks in terms of its impact on the discontinuity of ideas for innovation. We contrast the ideas in terms of their newness, their level of surprise and their requirement for new technologies to get close to their discontinues nature with regards to market and technological factors (Garcia & Calantone, 2002).

We thereby broaden the scope of user innovation research by incorporating insights from the micro processes of collaboration, focusing on the edges of markets – the potential users. Guidelines for how to use existing tools of user collaboration to generate discontinuous ideas are derived from this research. Our approach is interdisciplinary, combining the field of innovation management with creativity research. The introduction is followed by a literature review of the main concepts we want to explore in this study: the *user typology* and the *cognitive distance* between individuals in collaboration.

3.2 Theoretical Background

3.2.1 User Typology

The collaboration with users in innovation has been a well-respected field of study in recent years (e.g. Bharadwaj et al., 2012; Kristensson et al., 2004; Lettl et al., 2006; von

Hippel, 1986). The logic underlying these studies assumes that innovation inspired by users builds upon inherent and upcoming market needs and thereby has a higher probability of success (von Hippel, 1986). However, there has been an ongoing discussion on the role of users in the innovation process. Opponents of (current) user involvement argue that user research tends to undervalue radical innovations, and that any rational choice based on this research will lead to incrementalism (Christensen, 1997, 2006; Hamel & Prahalad, 1991; Markides, 2006; Ulwick, 2002; van der Panne et al., 2003). Therefore organizations should, they say, lead their users, not listen to them. However, user research should not be misinterpreted as a solution, but rather as an input for the innovation process. In order to create superior products, organizations have to understand their users' needs and build up institutional empathy (Leonard & Rayport, 1997; Slater & Mohr, 2006; Visser et al., 2007). Still, it seems plausible that participatory learning from current users might not be adequate, at least not exclusively, if new market opportunities are desired (Lynn, 1996).

Current User. We want to refer to two characteristics that might define current users and their contributions in value co-creation. The first characteristic relates to the tendency of current users to want to uphold internal consistency and continuity. Current users generally use a product or service to satisfy an established need and solve a problem. Therefore, current users tend to have little or no interest in deviating too far from established dominant designs and are less inclined to dispense with highly valued performance dimensions (Christensen, 1997). The behavior of the current user is described in consistency theories and their basic assumption that individuals have a need for consistency between attitudes and behaviors (e.g. Festinger, 1957; Rosenberg, 1956). A current user follows these personal objectives to prevent changes in valued product components when asked to imagine new and discontinuous products and services. Some social theories state that social properties can only change if their constituting individuals change (Sawyer, 2005). If an organization and its current users are seen as a social system, then it requires a change in behavior and perspectives on the part of current users in order to achieve transformation in dominant designs. Current users thereby exert dominance over the direction innovative activities might take towards more incremental paths (Fischer & Reuber, 2004). The second characteristic relates to the notion of functional fixedness in experienced people (e.g. Duncker, 1945). Current users are experienced users and learning from their experience is bound to the particular domain of their experience (Danneels, 2002; Levinthal & March, 1993). They accumulate experience by interaction with a product or service and are constrained by the mental models which emerge through interaction. It is a natural urge to think within the realm of known categories and schemes, even if one wants to diverge (e.g. Frensch & Sternberg, 1989; Koestler, 1964; Rubenson & Runco, 1995). Experience is different to expertise in the sense that it does not necessarily encompass professional training and conscious learning and might thereby result in less structured knowledge organization and representation (Chi et al., 1982). Kristensson et al. (2004) showed that ordinary users, lacking professional education within the respective domain, outperformed advanced users and organizational internal developers in idea generation in terms of the originality of ideas. This is the *tension view* of the relationship between creativity and knowledge (Weisberg, 1999). The *path of least resistance* model is even more radical, stating that once a known category of the domain at hand is accessible from memory, it will be used as the starting point for any creative generation task (Ward, 1994, 1995). This is especially valid for ill-defined problems, as there is no right or wrong at first (Voss & Post, 1998). People tend to reduce their cognitive effort when they reach a satisfying result, which is most easily derived from accessible categories. An opposing view, the *foundation view*, states that knowledge and experience provide the basis for creative engagement within a particular domain and are necessary preconditions for the achievement of significant advances within the domain (e.g. Bailin, 1988; Hayes, 1989). It is not clear whether deep immersion into or familiarity with a domain is preferable (Weisberg, 1999). This duality hints to the importance of cognitive distance in collaboration, which we will elaborate upon later within this article.

Potential User. If the focus of explorative activities is shifted towards the edges of markets, a company faces niche markets and potential users. Their early integration into the innovation process then becomes part of a deliberate scan for environmental changes outside the daily business of the organization (Danneels, 2003; Faems et al., 2005; Meyer, 1982). Chandy and Tellis (1998, p. 479) claim that potential users cause "decision makers in a firm to become keenly aware of market-related developments." It is a deliberate scanning of shifts in needs, social norms and values that define the boundaries of a market and its participants. Potential users might for example integrate technology into their environment in a novel fashion or relate their contextual experiences to the domain of the organization. In this way, new application domains or contexts might not cause considerable technological discontinuity, but rather discontinuity in user familiarity on the side of the organization (Danneels, 2002; Govindarajan et al., 2011). Potential users do not develop a clear expectation about how their needs - if existing within this domain - shall be addressed. Their needs might even be unfulfilled by the current dominant design and their contributions might therefore drive change in dominant designs. Their imagination is acknowledged to be flexible (Rubenson & Runco, 1995), as it is not constrained by experience from interaction with the products or services of a domain. From an information processing perspective, a potential user will relate abstract knowledge to a particular domain when confronted with a creative task (Alba & Hutchinson, 1987; Hewing, 2013; Ward, 1995). This abstraction has value, since it can broaden the solution space in unexpected, unusual ways in order to arrive at specific and original solutions. Its elusiveness also bears risks, however, which arise from problems of understanding at the recipients' side and from the postulate of the foundation view as explained earlier on (Weisberg, 1999). Potential users might struggle to advance the knowledge within a domain if they are not familiar with the current status quo. Consequently, an environment where these users can learn and acquire knowledge in a collaborative setting becomes all the more important.

3.2.2 Cognitive Distance in Collaboration

Exploration in organizations incorporates company-initiated semi-structured search processes that look for innovation impulses. These processes can be compared to ill-defined problem-solving processes (Leonard-Barton, 1995). Human problem-solving is constituted by the objective conditions of the problem, its subjective representation by the problem solver and the process by which these representations are manipulated (Duncker, 1945). The subjective representation of a problem is thereby strongly based on an individual's experience and the manipulation of representation on the ability for re-combination and flexibility. While the first refers to the experience of an individual within a particular domain, the second refers to the ability of an individual to somehow detach him- or herself from this experience to foster newness (Wiley, 1998). Often these two preconditions are not present within one individual (Rubenson & Runco, 1995). Collaborative processes can thus play an important role, bringing together individuals who represent at least one of these preconditions – experience or flexibility. Fertile manipulation of their individual representations can arise from mutual knowledge exchange and cognitive stimulation (Rubenson & Runco, 1995). Cognitive distance is a term used to describe, and an attempt to determine, the difference in cognitive structure between individuals, due to different experience paths (Nooteboom, 2000, 2009). When interacting, the difference in cognitive structure facilitates the chance to learn from the other's experiences, particularly in complex tasks. Therefore cognitive distance between individuals is imperative for learning and acquiring new knowledge, at least up to a certain threshold. Furthermore, emergent knowledge might arise through interaction, that is, knowledge that none of the people possessed a priori (Argote, 1999; Piaget, 1956). Emergent knowledge is a combination of previously unconnected entities (Sawyer, 2005; Ward, 2007). Yet cognitive distance as such is not sufficient. Collaborative efforts are needed to ensure that the unique knowledge held by each member is shared with the others. Research on collaboration has shown that members tend to share knowledge they have in common rather than proprietary information (Stasser & Titus, 1985), even if the common knowledge is little. This effect is called *convergence in collaboration* and shows the independence between the amount of shared knowledge and the sharing of knowledge between people (Hinsz et al., 1997). Innovation impulses often emerge in the interplay between knowledge bases (Paap & Katz, 2004; Reid & Bentrani, 2004), so it is important to understand under which conditions unique information is shared and understood. In user research, the processes for the creation, dissemination and utilization of knowledge in collaboration remain poorly understood (Bogers et al., 2010; Greer & Lei, 2012; Kristensson et al., 2008). Both potential and current users are a unique group of people, each of which follow different agendas, have individual contextual needs and different levels of explicit and sticky information about a domain (Hewing, 2013; von Hippel, 1994). We therefore deem it fruitful to bring potential and current users together to learn about the collaboration between and among them. We want to explore differences in the ideas between user typologies and between different constellations of cognitive distance. Thus no hypotheses are derived.

3.3   Method and Research Context

Reid and Bentrani (2004) argue that environmental insights for discontinuous ideas are brought into the organizational structure by key individuals who are involved in the market. This is supported by the notion that generative learning with users does not emerge through application of traditional market research methods, but rather by means of qualitative methods in which individuals immerse into users' lives (Henderson, 2006; Leonard & Rayport, 1997; Slater & Mohr, 2006; Visser et al., 2007). In line with this reasoning, we think that the merits of potential user collaboration are best researched at the point of their evolution, e.g. the outputs of participatory idea generation. We thus chose a quasi-experimental design to test the differences between potential and current users' ideas contributed after taking part in a dyadic idea generation discussion. Though psychology researchers emphasize the appropriateness of experimental approaches in research on creativity (Runco & Sakamoto, 1999; Sternberg & Lubart, 1999), it is a fairly underrepresented approach in research on innovation management. Experimental methods are well suited to the complex nature of creative problem-solving, due to the means of control and manipulation (Runco & Sakamoto, 1999). We examine creative collaboration in dyads in order to mitigate the effects of the social influences of groups (Rubenson & Runco, 1995).

The idea generation tasks were performed in the domain of music applications on mobile devices (e.g. mobile phones, tablets). Although there are many non-users within this domain, a large share will have some affiliation with music as such, and can thus be defined as potential users.

3.3.1   Experimental Constructs

Based on the theoretical discussion, three constructs are of central concern in our empirical analysis. The first is the user typology, represented by a current user (experienced) and a potential user (inexperienced). The second is the cognitive distance between collaborating individuals with regards to a domain, which is high between a potential and a current user and low between a dyad of potential users or a dyad of current users. Cognitive distance affects the potential to learn from one's dyadic partner, defines the level of shared and unique information and which information is shared in the collaboration. The third construct is the discontinuity of ideas. Garcia and Calantone (2002) framed discontinuous innovations to result in market and/or technological discontinuity on the level of a domain. Market discontinuities arise when organizations in the domain need to acquire new commercial competences, for example to serve unfulfilled needs. Technological discontinuities arise when the market context remains fairly stable, but new technological know-how or processes are needed. We therefore explore and compare the newness and surprising elements embedded in ideas of potential and current users, but also the need for new technologies to develop the ideas into future innovations.

3.3.2 Purposive Sampling

We recruited participants using a screening questionnaire to ensure the participation of both potential and current users and to check for influencing variables. The screening questionnaire was sent to university networks and to a user panel of the Telekom Innovation Laboratories. There was a participation incentive of ten Euro.

3.3.3 Pre-measures

The validated constructs *use experience* (Alba & Hutchinson, 1987) and *involvement* (Zaichkowsky, 1985) were drawn upon as criteria to classify each of the 142 respondents to the screening questionnaire as a current user, a potential user or as not suitable for the experiment. All constructs and their items can be found within the Appendix of this article.

Use experience. The construct use experience was used to account for the user typology and is a classificatory variable. It represents knowledge through direct interaction between a product and its user. Cognitive structure and mental models originate from this interactive process. This structure provides the basis for further involvement with the domain (Alba & Hutchinson, 1987). Respondents who used one or more music applications on a mobile device on a weekly basis were classified as current users. Those who did not use any application - whether music applications or other - on a mobile device were classified as potential users, in cases where their level of involvement matched.

Involvement. We used an adapted involvement construct (Chronbach's alpha = .79) to ensure that potential users could connect with the context of the domain. The average involvement construct of a potential user was expected to exceed the value of two in order to exclude respondents with no interest in the domain and to comply with the former definition.

Within the two classes, potential and current users, 34 participants of each class were selected and assigned randomly to low or high cognitive distance dyads.

User knowledge. The adapted user knowledge construct (alpha = .86; Park et al., 1994) was incorporated to detect differences in levels of knowledge between current and potential users. Use experience and user knowledge are not independent of each other, but they nevertheless relate to different types of knowledge (Alba & Hutchinson, 1987; Park et al., 1994). User knowledge represents the body of knowledge on a broader scope. It encompasses sources other than usage and might be classified as more theoretical expertise (Brucks, 1985) and therefore used as a manipulation check of the user classification within our sample.

Personality inventory scales. The above-mentioned constructs are domain-dependent, which means that they measure a relation between a person and the domain. The screening questionnaire also contained adapted domain-independent constructs such as *empathy* (alpha = .78; Zeidner et al., 2004) and *creativity* (alpha = .83; Kirton, 1976; Im et al., 2003), which are personal inventory constructs. Empathy is an affective disposition on the part of an individual that has the potential to enhance collaborative outcome (Zeidner et al., 2004). It is reflected in the awareness of, and response to, the feelings of others and is not only important to understand the reasoning of others, but also to keep a collaborative process alive. As a further precaution, we added two questions regarding the participant's frequency in engaging in and education in creative problem-solving, to check for influences coming from an individual's expertise in performing problem-solving tasks (Argote & Kane, 2003). We added and averaged the values and refer to them as *problem-solving expertise* (alpha = .73) from here on.

Table 3.1: Occupation of the Sample (n = 68)

	Cases	%
Student	48	70.6
Employee	11	16.1
Self-Employed	5	7.4
Trainee	2	2.9
Unemployed	1	1.5
Other	1	1.5

The above personality inventory constructs serve as control variables. The respective occupations of the participants are listed within Table 3.1. In total 68 participants were invited to discuss their ideas on music applications on mobile devices in dyads, and to subsequently sketch one idea individually.

3.3.4 Study Design and Instruments

We chose a 2x2 between-subject factorial block design to look at systematic differences in the discontinuity of the ideas between the different experimental units. In total, we conducted 34 dyadic idea discussions within one week. After the dyadic idea discussions each of the participants was asked to specify one idea individually on an idea template, resulting in 68 ideas. Four experimental units arose through the treatment: an idea that was contributed by (1) a potential user facing a potential user before, (2) a potential user facing a current user before, (3) a current user facing a potential user before and (4) a current user facing a current user before, as depicted in Table 3.2.

The experiment went through the following successive steps:

Start-up. The first step included an introduction of the participants and a brief introduction to the idea discussion.

Table 3.2: Factorial Research Design

		Main Effect 2: Cognitive Distance	
		low	high
Main Effect 1: User Typology	Current User	n = 18 (2/16)	n = 16 (6/10)
	Potential User	n = 18 (7/11)	n = 16 (4/12)

(female/male)

Warm-up. The participants were asked to perform two so-called icebreaker tasks. The first task was designed to enable them to become more familiar with each other and to reduce the barrier between researcher and participants. Trust and cohesiveness through informal communication before creative tasks is conducive for its outcome (Edmondson, 1999). The second one was a short collaborative creativity task to help participants lose fear they might have of talking out loud, become comfortable to articulate their thoughts and get an idea of how to proceed in the main task.

Idea discussion. The main creative problem-solving task was to discuss new potential music applications for mobile devices. It was read and handed out to the participants, so they could look at it again while discussing. The participants were told to interpret *new* as being *new to them.* Hence it was an open-ended problem-solving task, which allowed individuals to be creative and engage in divergent thinking (Guilford, 1968; Runco & Sakamoto, 1999). They were asked to verbalize as many of their thoughts as possible to their discussion partner, since sharing ideas is an important component of collaborative and creative problem-solving (Paulus & Yang, 2000). This step lasted 20 minutes. The researcher was in the room, but silent and withdrawn from the space where the two participants were working, so that they would not feel observed and would not seek the researcher's response (e.g. reassurance or guidance).

Idea template. After the idea discussion the participants were separated and each was provided with an idea template. The task was to specify one idea in detail and written form, to give the idea a name and to describe its purpose. Specifying one idea individually gave each participant the chance to reflect on the ideas that had come up in the course of their interaction. Individual reflection is important for collaboration to have a positive effect on the outcome (Paulus & Yang, 2000). Also, it allows one to check for cross-fertilization effects between different levels of cognitive distance separately.

Debriefing. The experiment concluded with a short debriefing of the participants.

3.3.5 Evaluation

Two months after the idea discussions, an expert panel evaluated each idea template based on the Consensual Assessment Technique (Amabile, 1996). In cases where more than one participant had specified the same idea, the idea was presented only once to the expert panel and each participant's idea obtained the same evaluation score. This procedure not only simplified the task of the experts to some extent, but also ensured that different wording of the same idea would not affect its performance. Two researchers were required to agree that a particular idea was identical to another idea before they were consolidated, leaving a number of 61 unique ideas for evaluation[3]. Out of the 61 unique ideas only one contained a reference to the name of an existing solution, which was eliminated from the wording in order not to bias the expert's evaluation. For the assessment of the ideas, three criteria were selected: *newness* (Amabile, 1996), *surprise* (Jackson & Messick, 1965) and the *requirement for new technologies* (Goldenberg et al., 2001). The original criteria (newness; value; producibility) were adapted to account for the discontinuity of ideas to dominant designs of the market. The adaptation was based on previously used criteria (surprise and requirement for new technologies) and on the separation of discontinuity into market and technological discontinuity (Garcia & Calantone, 2002). Newness alone is not an expression of discontinuity, since it can also reflect a very original but somewhat continuous idea (e.g. Audia & Goncalo, 2007; Sternberg et al., 2003). To complement newness, we added surprise. The notion of

---

[3] Note that identical ideas received the same scores assigned. A total of 68 data units are therefore integrated into the subsequent analysis.

surprise stands for unusualness in a product or service, judged against dominant market designs, and thereby relates to market discontinuity (Amabile, 1996; Jackson & Messick, 1965). These unusual cues might refer to emerging ideas or provide a fresh insight for the expert embedded in the ideas that can be combined with established product knowledge to foster market discontinuities (Garcia & Calantone, 2002; Quinn, 1985; Ward, 2007). Consequently, an idea that is perceived as highly novel and surprising constitutes potential for high market discontinuity. We used requirements for new technologies (Goldenberg et al., 2001; Veryzer, 1998) instead of the often-used producibility criterion, since we were particularly interested in the effects of our experimental factors on the technological competences of an organization within the respective domain. This relates to whether ideas, which emerge from the edges of a domain, require new technological competences or whether existing competences can be expanded over established domain boundaries.

The expert panel consisted of six professionals from the music and music software industry, two of whom had backgrounds in technical engineering, two came from the realm of marketing and another two from business development. Since the ideas were to be evaluated relatively to each other, a sample of ten randomly picked ideas had to be read by each expert before evaluation in order to get a picture of the variety of ideas. The experts rated the idea templates independently of each other. All six experts rated each of the 61 unique ideas on all three criteria on a seven-point Likert scale, from one (very low) to seven (very high).

In order to make evaluation and understanding of potentially discontinuous music applications more explicit, two exemplary ideas are presented here – one that was rated as rather discontinuous (Example 1) and one that was rated as rather continuous (Example 2).

Example 1. With this app you can explore a city music-wise. When you pass by an important sight in the history of music (e.g. a famous recording studio), the app would play back a song from that area and give you additional audio-visual information about the musicians. In the absence of sights, it can show you what kind of music is typical for that region or what music is frequently consumed there.

Example 2. *This app helps you to identify and play new music. You hold your mobile device towards a music speaker, record a fraction of the song that is being played and it will not only tell you the name of the song and artist, but will also display the lyrics or instrumental sheets in real time.*

The reasons for the evaluations are implied within the scores provided by the experts, but in this example, the first idea certainly addresses a market (tourism) that is different to the common market in the music industry. In contrast, the second example is an extension of features that already existed within music applications at the time the experiment was conducted.

3.4   Results

3.4.1   Reliability of Evaluation

We used the intraclass-correlation-coefficient (ICC) to measure degrees of consensus among the six expert judges (McGraw & Wong, 1996; Shrout & Fleiss, 1979). Values equal to or above .7 are considered to reflect high consensus. As depicted in Table 3.3, all evaluations met this threshold.

Table 3.3: Intraclass-Correlations-Coefficients of Evaluation Criteria

Newness	Surprise	Requirement for New Technologies
.71	.70	.70

3.4.2   Control Variables

We ran an independent t-test on the user knowledge construct to ascertain differences in knowledge between potential and current users (manipulation check). On average, there is a significant difference in user knowledge (potential users $M = 2.59$, $SE = .13$; current users $M = 3.24$, $SE = .14$) with $t(66) = 3.33$ and $p < .05$.

Since there is no significant correlation between the dependent and the control variables, they do not meet the necessary conditions to be incorporated as covariates into the overall model (Hair et al., 2006; Tabachnick & Fidell, 2007). Instead, we ran a separate one-way analysis of variance (ANOVA) with Bonferroni post-hoc tests to assess statistical significance. None of the experimental units showed a significant difference with regards to the constructs creativity, empathy, involvement, problem-solving expertise or age. There was a disproportion between female and male participants (see Table 3.2). We therefore tested differences in the evaluation criteria of the ideas according to gender, with no significant differences in any of the criteria.

3.4.3 Dependent Variables

The effects of the experimental factors user typology and cognitive distance on the newness, surprise and requirement for new technologies of an idea are tested by means of a factorial analysis of variance (ANOVA).

Newness. Both main effects, user typology and the cognitive distance, appeared to have a non-significant effect on newness, but there was a significant interaction effect between them, with $F(1,64) = 6.2$ and $p < .05$. Thus, the main effects become less informative, since the interaction term indicates that potential and current users were affected differently by cognitive distance, as depicted in Figure 3.1.

Figure 3.1: Simple Slope Analysis for Newness

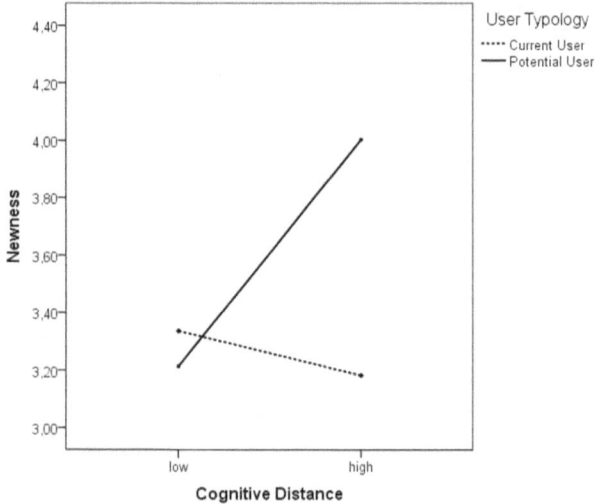

Since the interaction term does not tell us in what way the experimental units differ, we ran various simple effect analyses. Within the high cognitive distance dyads, there was a significant difference in the newness of ideas between potential users ($M = 4.00$; $SE = .16$) and current users ($M = 3.18$; $SE = .25$), with $F(1,65) = 8.61$, $p < .01$. Within the potential user class, there was a significant difference in the newness of ideas between potential users from high cognitive distance dyads ($M = 4.00$; $SE = .16$) and potential users from low cognitive distance dyads ($M = 3.21$; $SE = .18$), with $F(1,65) = 7.95$, $p < .01$.

Surprise. There was a significant main effect of the user typology on the surprise of ideas in favor of the potential users, with $F(1,64) = 11.4$ and $p < .01$. The main effect of cognitive distance was not significant, neither was the interaction effect between them. Still it is interesting to take a look at the simple effect analysis. Within the potential user class, there was a significant difference in the surprise of ideas between potential users from high cognitive distance dyads ($M = 4.33$; $SE = .21$) and potential users from low cognitive distance dyads ($M = 3.61$; $SE = .21$), with $F(1,65) = 4.20$, $p < .05$.

Figure 3.2: Simple Slope Analysis for Surprise

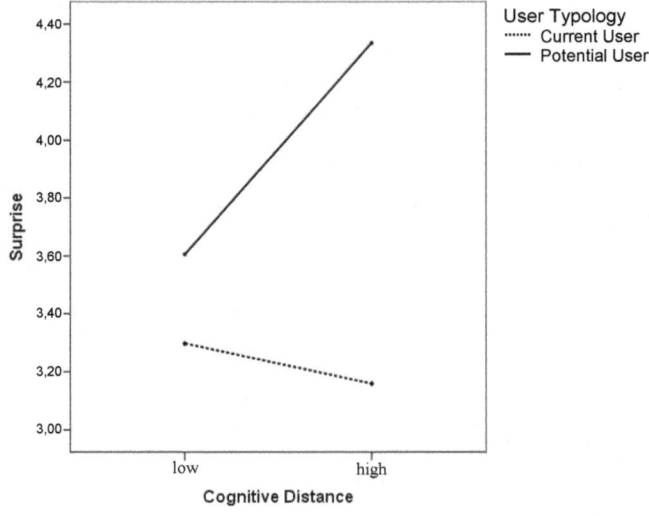

Though the main effect user typology was significant, it becomes obvious that the remarkably high evaluation of the ideas from potential users from high cognitive dyads was mainly responsible for this result.

Requirement for New Technologies. None of the main effects, neither the interaction effect nor the simple effect analyses, showed a significant effect on the technological discontinuity of ideas, as depicted in Figure 3.3.

Figure 3.3: Simple Slope Analysis for Requirements for New Technologies

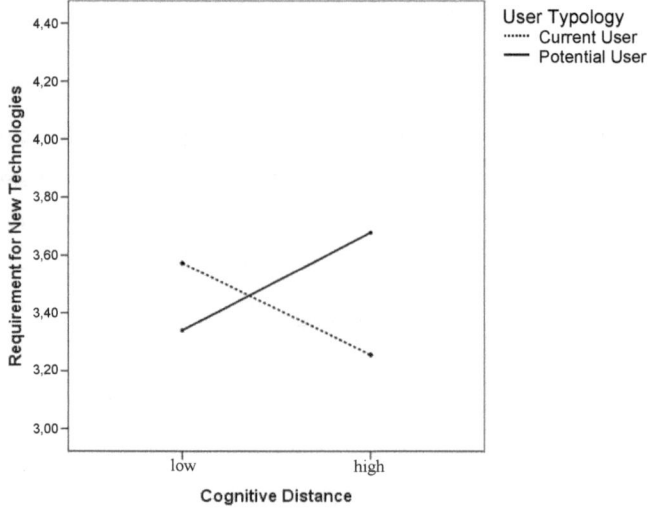

Table 3.4: Mean Values of the Evaluation Criteria

	Potential User		Current User	
Cognitive Distance	low	high	low	high
Newness	3.21	4.00** / *	3.34	3.18
Surprise	3.61	4.33** / **	3.30	3.16
Requirement for New Technologies	3.34	3.68	3.57	3.26

* $p < .05$; **$p < .01$; (simple effect between potential and current user from high cognitive distance dyad) / (simple effect between high and low cognitive distance dyads within the potential user class)

To sum it up, potential users from high cognitive distance dyads contributed ideas that were significantly higher in newness and surprise compared to any other experimental unit. Table 3.4 shows the mean values of each evaluation criteria and the significance levels of the simple effect analyses. Especially the criterion of surprise sets the potential users from high cognitive distance dyads apart from all others, with an effect size[4] of r_{surp} =.42, r_{new} =.39 and r_{tech} = .18.

3.5 Discussion and Conclusion

In this research, the merits of involving users into participatory creative problem-solving and idea generation have been analyzed with a focus on including potential users in varying collaborative compositions. The results of a quasi-experimental design highlight the idea that potential users are valuable external partners when it comes to exploring new market opportunities. They are able to contribute ideas that are more discontinuous in nature compared to current users. This is especially the case when they have the opportunity to collaborate with experienced people. It shows the importance of cognitive distance in participative approaches to creative problem-solving with users. Current users provide structural alignment to the problem at hand that is absorbed by potential users, but only as much as needed.

The criterion of surprise distinguishes the ideas of potential users from high cognitive distance dyads most obviously from the others. The expert evaluators involved in this research emphasized the unusual, emergent and combinatorial properties of an idea these people brought forward, properties that might upset established practice in a domain (Christensen & Bower, 1996). The unusualness in the ideas of potential users might arise from unique information and their effort to share and explicate their tacit knowledge. They also absorb the domain-related knowledge of current users and relate it to their own imagination and innovativeness. This notion is underpinned when looking at the impact on the technological requirements of their ideas and the poorer performance of the low cognitive dyads. Though there was a correlation between newness, surprise and requirements for new technologies, there were no significant differences in the

---

[4] In experimental research, the Pearson's correlation coefficient r is frequently used to report effect sizes (Field, 2001).

technological discontinuity of ideas between any of the experimental units. This indicates that ideas from potential users do not deliberately call for technological changes, but rather yield leaps in customer familiarity, e.g. new needs and application contexts, that can be addressed with rather slight technological adaptations. This complies with the definition of really new innovations by Garcia and Calantone (2002) that yield either market or technological discontinuities, but not both. New application contexts, for instance, call for dynamic adaptation of an organization's capabilities in terms of their understanding of these contexts and the involved individuals. In line with the competence based typology of innovations (Danneels, 2002), organizations can learn how to leverage their technological competences by delinking these from established products and linking them to new social contexts when collaborating with potential users.

As stated above, potential users' ideas are most discontinuous if they collaborate with a current user. The simple effect analysis showed the most severe differences in newness and surprise of the ideas in comparisons between potential and current users within high cognitive distance dyads. This reveals a limitation of the notion of cognitive distance applied to user research, especially within the current user class. Current users' ideas are not significantly affected by cognitive distance in a participatory setting. The cross-fertilization and exchange of cognitive stimuli inherent in collaborative processes seem to be asymmetric in their impact on the outcome of the creative task with these different user types. The potential user benefits from the reference frame transferred by the current user through interaction, while the current user does not absorb the stimuli coming from the potential user which might be more distant to known routines. Two theories might explain this observation.

Sticky information. First, the potential user is obliged to rely on and to share uniquely held knowledge with the discussion partner. This knowledge is more difficult to transfer than domain-related knowledge of the current user as it is more sticky and tacit in nature (Polanyi, 1966; von Hippel, 1994). It requires expertise to identify, understand and synthesize tacit knowledge. A current user might lack this expertise and therefore will have problems to follow the potential user's reasoning. Current users' domain related knowledge concerning solutions and the way they work in contrast could easily be made explicit, even to an inexperienced person. The potential user can use and connect this domain-related knowledge to his or her own imagination and inventiveness. Therefore, in order to learn from each other, different levels of cognitive effort and skill are required from a potential and a current user.

Consistency. Second, even if a current user identifies and understands the contextual cues a potential user is trying to verbalize, these might seem trivial, irrelevant or even in opposition to his or her goals. Current users might have an interest in sustainability of dominant designs (Christensen 1997, 2006; Henderson, 2006), as postulated in consistency theories (e.g. Festinger, 1957; Rosenberg, 1956). They are utility maximizers and use a product and certain components of it for a particular purpose (Bogers et al., 2010). Their approach to new ideas is more problem-oriented with regard to established solutions. This leads current users to refuse compromises on crucial dimensions of established products, which is detrimental if discontinuity is the goal. They seek harmonization between their attitude and their behavior in creative problem-solving. Here, their personal objective collides with the task objective (Hinsz et al., 1997). A potential user can therefore not only learn more easily, but might also be more willing to do so than vice versa.

Potential users who collaborate with equally inexperienced users yield significantly lower ideas with regards to newness and surprise compared to potential users in high cognitive distance dyads. As new ideas emerge from roots within established knowledge (Bailin, 1988), collaboration between inexperienced people lacks a frame of reference and a chance for mutual learning (Hayes, 1989). In other words, potential users struggle to advance the knowledge within a domain if they are not aware of already existing knowledge. Yet this might not only be due to missing structural alignment, but especially due to the participatory effect. Potential users need to not only cope with their own uncertainty, but also with that of their respective discussion partner. This might lead to inefficient processes of idea generation and uncertainty (Weisberg, 1999).

Current users from low cognitive distance dyads encounter two disadvantages over potential users from high cognitive distance dyads. First, they are subject to effects postulated by the path of least resistance model and by the functional fixedness described above. Second, as the cognitive distance to their collaborative partner is marginal, their chances to learn are relatively slim. Furthermore, research on information sampling suggests that members of collaborations are more likely to express shared perspectives during discussions than unique ones (Hinsz et al., 1997; Stasser, 1992). Consequently, both current users tend to share and conform on the domain-related knowledge they both possess and thereby arrive at predictable solutions.

From a methodological perspective, there is a backlog of experimental research on creative problem-solving that analyzes the impact of manipulation of explicit knowledge (Runco and Sakamoto, 1999). The present study examines how tacit knowledge affects the process. This is a first step in strengthening collaborative experimental research within the domain of user research.

To sum up, our analysis of potential and current user's ideas in participatory creative problem-solving tasks yields three interesting insights which define the merits of user integration more precisely when increased discontinuity of the outcome is desired. Firstly, potential users contribute ideas that are perceived more original (newer) than those of current users, especially when they collaborate with someone experienced within this domain. Secondly, a potential user's ideas are most obviously characterized by an increased level of surprise that seems to arise from emergent properties embedded in the ideas or new contextual relations and does not require technological leaps. Lastly and maybe most interestingly, the effectiveness of cross-fertilization between two cognitive distant users is asymmetric, meaning that potential users benefit more (or are more willing to benefit) from the current users' experience than the current user from the flexibility of the potential user.

3.5.1 Implications for Practice

We showed that the potential user is an attractive partner within explorative search processes when certain methodological preconditions are met. Thus our study reveals limitations to the notion of involving potential users in explorative processes, which arise from the micro perspective approach. Though ideas might only constitute an intermediate step on the road to future innovation, they are a more precise subject of analysis to understand users' contributions and behavior. Organizational search for better processes at the early idea stage can benefit from these findings. The potential users' salient contributions and impact in terms of discontinuity in participatory processes do need structural alignment to the dominant designs. The potential user needs cognitive stimuli through constant response over the creative process. Furthermore, our findings reveal limits to the notion of cross-fertilization from cognitive distance in so far as current users do not benefit from the exchange in the same way. These findings render the potential user notion and its merits in terms of content and impact more precisely, making it more concrete and tangible. Neither user class (potential or current) on its own improves the process in collaborative settings; both parties are needed for an effective process. In these settings, key individuals of the organization can acquire new market capabilities that will be a valuable asset in the later process. Attention should focus thereby on the contributions by potential users who are in exchange with someone experienced. Their contributions are more surprising and unusual, which can also mean that they are unsettling. This provocation might initiate a reassessment of one's current way of doing, one's market position and future vision. The tacit knowledge of the potential user can provide the organization with a competitive advantage. Collaboration with potential users therefore relates to contextual approaches of real-time and short-term learning within research on organizational learning (Miner et al., 2001).

Furthermore, a subsequent individual reflection of the interaction and combined ideas should be assured, since the two individuals follow different agendas. Current users in this case tend to implicitly follow an agenda with a focus that is embedded in the dominant design. Both idea outputs are useful and can depict a roadmap from more obvious improvements of dominant designs to more discontinuous projects. Also, if the merits of potential user collaboration are difficult to absorb, their involvement at later stages in the innovation process might be fruitful in order not to lose the initial impulse they were coming from. When viewed from an organizational perspective, the integration of potential users can provide a competitive edge in growing markets, while it is not detrimental to sustaining and serving the current market.

3.5.2 *Limitations and Future Research*

The scope of this research is on the early idea stage in innovation processes, which is conducted with users. Some limitations of this study should be emphasized, which also provide further research opportunities. The sample size is fairly small with 68 individuals, especially when it comes to the simple effect comparisons of the varying factorial levels. Glaser (1978) states, that at least 10 observations per class are needed, Hair et al. (2006) recommend 20 observations. The main effects comply with both propositions, while the classes for interaction effects at least comply with the former. Though block designs generally reduce error variance (Liebert & Liebert, 1995), increasing the sample size would yield more robust results with regards to chance factors. In future research, an individual control group could be set up including users, who not collaborated, but developed ideas individually. An individual control group sheds light on whether a dyad of potential users results in a negative effect of mutual uncertainty and would be outperformed by an inexperienced individual.

As this is an experimental research, the internal and external validity of the findings should be addressed. Several precautions were taken to ensure internal validity, such as using a homogenous sample in terms of age and occupation (mostly students) and controlling for potential influencing variables such as creativity, empathy, expertise in creative problem-solving and involvement. The latter are self-reported measures, which might have the inherent method effect of social desirability (Conway & Lance, 2010). However, the target person itself is the only one privileged to access this information in this case where no outside source is available. Furthermore, anonymity of the data was explicitly assured, which reduces social desirable responding (Heneman, 1974). The homogeneity of the sample with regards to age and occupation compromises external population validity. One intriguing finding is that the cross-fertilization in cognitive distant dyads seems to be asymmetric in its direction due to the difference in the nature of exchanged knowledge. First of all, this finding should be substantiated in a qualitative setting. Furthermore, it certainly might vary with the particular domain and its inherent complexity. Similar studies in various domains, also ones that have a rather negative emotional impact on people (e.g. dealing with serious health problems) would be an insightful research path. External validity of our findings is also limited to the selected method of user involvement, which was an idea generation task. Comparing insights from other methods, especially coming from ethnography (e.g. shadowing) is definitely of interest within the environmental search of organizations.

Idea discussions were conducted between two users with similar or distinct preconditions. It might be fruitful to extend the research to the corporate level, where a member of the company discusses ideas with users, potential and current ones. Firstly, their expertise in identifying and synthesizing environmental stimuli and sticky information might mitigate the asymmetric knowledge exchange. Secondly, they bring in different knowledge such as what kind of ideas go well within the organization or what kind of ideas has been rewarded in the past (Sternberg, 1998). This procedural knowledge is seen to directly or unconsciously affect the creative problem-solving process. Furthermore, contrasting ideas of lead and potential users deems insightful, since integrating potential users in the context of new market exploration might not always be the dominant strategy. At this stage, we believe the value of collaboration with potential users is greatest at the early stage of idea generation. In contrast, lead users also play an active part in the actual conception and prototyping of innovations (Lüthje, 2004; von Hippel, 1986). Their impact has far-ranging influence on the implementation phase. However, creativity research informs us that it is not always the deeply knowledgeable and interwoven person who presents the ability to radically change a domain (e.g. Duncker, 1945; Rubenson & Runco, 1995; Weisberg, 1999). This is particularly relevant when the goal is to redefine the boundaries of a domain through explorative learning (Lettl, 2007).

In conclusion, we have shown the merits of potential user collaboration and under what circumstances the impulses of potential users can be fruitfully explored. This enhances the knowledge of user typologies and has methodological implications for the early stages in innovative processes. The approach might also serve as a basis for further empirical studies that involve the edges of markets.

References

Adner, R. 2002. When are Technologies Disruptive? A Demand-based View of the Emergence of Competition. *Strategic Management Journal,* 23(8): 667–88.

Alba, J. W., & Hutchinson, W. J. 1987. Dimensions of Consumer Expertise. *Journal of Consumer Research,* 13(4): 411–454.

Amabile, T. M. 1996. *Creativity in Context: Update to the Social Psychology of Creativity.* Boulder, CO: Westview Press.

Argote, L. 1999. *Organizational Learning: Creating, retaining and transferring knowledge.* Norwell, MA: Kluver Academic.

Argote, L., & Kane, A. A. 2003. Learning from Direct and Indirect Experience in Organizations: The Effects of Experience Content, Timing, and Distribution. In P. B. Paulus & B. A. Nijstad (Eds.), *Group Creativity:* 277-303. Oxford University Press.

Arnold, T. J., Fang, E., & Palmatier, R. W. 2011. The Effects of Customer Acquisition and Retention Orientations on a Firm's Radical and Incremental Innovation Performance. *Journal of the Academy of Marketing Science,* 39(2): 234-251.

Audia, P. G., & Goncalo, J. A. 2007. Past Success and Creativity over Time: A Study of Inventors in the Hard Disk Drive Industry. *Management Science,* 53(1): 1-15.

Bailin, S. 1988. *Achieving Extraordinary Ends: An Essay on Creativity.* Dodrecht: Kluwer Academic.

Bharadwaj, N., Nevin, J. R., & Wallman, J. P. 2012. Explicating Hearing the Voice of the Customer as a Manifestation of Customer Focus and Assessing its Consequences. *Journal of Product Innovation Management,* forthcoming.

Birkinshaw, J., Bessant, J., & Delbridge, R. 2007. Finding, Forming, and Performing: Creating Networks for Discontinuous Innovation. *California Management Review,* 49(3): 67-84.

Bogers, M., Afuah, A., & Bastian, B. 2010. Users as Innovators: A Review, Critique, and Future Research Directions. *Journal of Management,* 36: 857–875.

Bonner, J. M, & Walker Jr., O. C. 2004. Selecting Influential Business-to-Business Customers in New Product Development: Relational Embeddedness and Knowledge Heterogeneity Considerations. *Journal of Product Innovation Management,* 21: 155-169.

Brucks, M. 1985. The Effects of Product Class Knowledge on Information Search Behavior. *The Journal of Consumer Research,* 12: 1-16.

Chandy, R. K., & Tellis, G. J. 1998. Organizing for Radical Product Innovation: the Overlooked Role of Willingness to Cannibalize. *Journal of Marketing Research,* 35(4): 474–487.

Chi, M., Glaser, R., & Rees, E. 1982. Expertise in Problem-solving. In R. S. Sternberg (Ed.), *Advances in the Psychology of Human Intelligence,* vol. 1: 1-75. Hillsdale, NJ Erlbaum.

Christensen, C. M. 1997. *The Innovator's Dilemma.* Boston: Harvard Business School Press.

Christensen, C. M. 2006. The Ongoing Process of Building a Theory of Disruption. *Journal of Product Innovation Management*, 23: 39-55.

Christensen C. M., & Bower, J. L. 1996. Customer Power, Strategic Investment, and the Failure of Leading Firms. *Strategic Management Journal*, 17(3): 197–218.

Conway, J. M., & Lance, C. E. 2010. What Reviewers Should Expect from Authors Regarding Common Method Bias in Organizational Research. *Journal of Business and Psychology*, 25:325–334

Crossan, M. M., Lane, H. W., & White, R. E. 1999. An Organizational Learning Framework: From Intuition to Institution. *Academy of Management Review*, 24(3): 522–537.

Danneels, E. 2002. The Dynamics of Product Innovation and Firm Competences. *Strategic Management Journal*, 23: 1095-1121.

Danneels, E. 2003. Tight-Loose Coupling with Customers: The Enactment of Customer Orientation. *Strategic Management Journal*, 24: 559-576.

Day, G. S. 1999. Misconceptions about Market Orientation. *Journal of Market Focused Management*, 4: 5-16.

Duncker, K. 1945. On Problem-solving. *Psychological Monographs*, 58(5): whole no. 270.

Edmondson, A. 1999. Psychological Safety and Learning Behavior in Work Teams. *Administrative Science Quarterly*, 44(2): 350-383.

Faems, D., van Looy, B., & Debackere, K. 2005. Interorganizational Collaboration and Innovation: Toward a Portfolio Approach. *Journal of Product Innovation Management*, 22: 238–250.

Festinger, L. 1957. *A Theory of Cognitive Dissonance.* Stanford, CA: Stanford University Press.

Field, A. P. 2001. Meta-analysis of Correlation Coefficients: a Monte Carlo Comparison of Fixed- and Random-effects Methods. *Psychological Methods*, 6: 161-180.

Frensch, P. A., & Sternberg, R. J. 1989. Expertise and Intelligent Thinking: When is it Worse to Know Better? In R. J. Sternberg (Ed.), *Advances in the Psychology of Human Intelligence*, vol. 5: 157-158. Hillsdale, NJ Erlbaum.

Fuchs, C., & Schreier, M. 2011. Customer Empowerment in New Product Development. *Journal of Product Innovation Management*, 28(1): 17-32.

Garcia, R., & Calantone, R. 2002. A Critical Look at Technological Innovation Typology and Innovativeness Terminology: A Literature Review. *Journal of Product Innovation Management*, 19(2): 110–132.

Glaser, W.R. 1978. *Varianzanalyse.* Stuttgart.

Goldenberg, J., Lehmann, D. R., & Mazursky, D. 2001. The Idea Itself and the Circumstances of Its Emergence as Predictors of New Product Success. *Management Science*, 47: 69-84.

Govindarajan, V., & Kopalle, P. K. 2006a. Disruptiveness of Innovations: Measurement and an Assessment of Reliability and Validity. *Strategic Management Journal*, 27: 189-199.

Govindarajan, V., & Kopalle, P. K. 2006b. The Usefulness of Measuring Disruptiveness of Innovations Ex Post in Making Ex Ante Predictions. *Journal of Product Innovation Management*, 23: 12-18.

Govindarajan, V., Kopalle, P. K., & Danneels, E. 2011. The Effects of Mainstream and Emerging Customer Orientations on Radical and Disruptive Innovations. *Journal of Product Innovation Management*, 28(S1): 121-132.

Greer, C. R., & Lei, D. 2012. Collaborative Innovation with Customers: A Review of The Literature and Suggestions for Future Research. *International Journal of Management Reviews*, 14(1): 63–84.

Guilford, J. P. 1968. *Creativity, Intelligence, and their Educational Implications.* San Diego, CA.

Hair, J. F., Black, W. C., Babin, B. J., Anderson, R. E., & Tatham, R. L. 2006. *Multivariate data analysis.* Upper Saddle River.

Hamel, G., & Prahalad, C. K. 1991. Corporate Imagination and Expeditionary Marketing. *Harvard Business Review*, 69(4): 81–92.

Hang, C., Chen, J., & Subramian A. M. 2010. Developing Disruptive Products for Emerging Economies: Lessons from Asian cases. *Research Technology Management:* 21-26.

Hauser, J. R., Tellis, G. J., & Griffin, A. 2006. Research on Innovation: A Review and Agenda for Marketing Science. *Marketing Science*, 25(6): 687-717.

Hayes, J. R. 1989. Cognitive Processes in Creativity. In J. A. Glover, R. R. Ronning & C. R. Reynolds (Eds.), *Handbook of Creativity:* 135-145. New York: Plenum.

Henderson, R. 2006. The Innovator's Dilemma as a Problem of Organizational Competence. *Journal of Product Innovation Management*, 23: 5-11.

Heneman, H. G. 1974. Comparisons of Self- and Superior Ratings of Managerial Performance. *Journal of Applied Psychology*, 59: 638–642.

Hewing, M. 2013. Merits of Collaboration with Potential and Current Users in Creative Problem-Solving. *International Journal of Innovation Management*, 17(3). 22.

Hinsz, V. B., Tindale, R. S., & Vollrath, D. A. 1997. The Emerging Conceptualization of Groups as Information Processors. *Psychological Bulletin*, 121(1): 43-64.

Hoffman, D. L., Kopalle, P. K., & Novak, T. P. 2010. The Right Consumers for Better Concepts: Identifying and Using Consumers High in Emergent Nature to Further Develop New Product Concepts. *Journal of Marketing Research*, 47: 854-865.

Im, S., Bayus, B. L., & Mason, C.H. 2003. An Empirical Study of Innate Consumer Innovativeness, Personal Characteristics, and New-Product Adoption Behavior. *Journal of the Academy of Marketing Science*, 31: 61–73.

Jackson, P., & Messick, S. 1965. The Person, the Product and the Response: Conceptual Problems in the Assessment of Creativity. *Journal of Personality*, 33: 309-329.

Kirton, M. 1976. Adaptors and Innovators: A Description and Measure. *Journal of Applied Psychology*, 61(5): 622–29.

Koestler, A. 1964. *The Act of Creation.* London: Picador.

Kristensson, P., Gustafsson, A., & Archer, T. 2004. Harnessing the Creative Potential among Users. *Journal of Product Innovation Management,* 21: 4-14.

Kristensson, P., Matthing, J., & Johansson, N. 2008. Key Strategies for the Successful Involvement of Customers in the Co-Creation of New Technology-Based Services. *International Journal of Service Industry Management,* 19(3-4): 474-491.

Lau, A. K. W., Tang, E., & Yam, R. C. M. 2010. Effects of Supplier and Customer Integration on Product Innovation and Performance: Empirical Evidence in Hong Kong Manufacturers. *Journal of Product Innovation Management,* 27(5): 761-777.

Leonard-Barton, D. 1992. Core Competencies and Core Rigidities: a Paradox in Managing New Product Development. *Strategic Management Journal,* 13: 111–125.

Leonard-Barton, D. 1995. *Wellsprings of Knowledge.* Boston: Harvard Business School Press.

Leonard, D., & Rayport, J. F. 1997. Spark Innovation through Empathic Design. *Harvard Business Review,* 75: 102-113.

Lettl, C. 2007. User Involvement Competence for Radical Innovation. *Journal of Engineering and Technology Management,* 24: 53-75.

Lettl, C., Herstatt, C., & Gemünden, H. G. 2006. Users' Contributions to Radical Innovation: Evidence from Four Cases in the Field of Medical Equipment Technology. *R&D Management,* 36: 251-272.

Levine, J. M., Resnick, L. B., & Higgins, E. T. 1993. Social Foundations of Cognition. *Annual Review of Psychology,* 44: 585-612.

Levinthal, D. A., & March, J.G. 1993. The Myopia of Learning. *Strategic Management Journal,* 14: 95–112.

Liebert, R. M., & Liebert, L. L. 1995. *Science and Behavior: an Introduction to Methods of Psychological Research.* Englewood Cliffs, NJ: Prentice Hall.

Lüthje, C. 2004. Characteristics of Innovating Users in a Consumer Goods Field. *Technovation,* 24(9): 1-28.

Lynn, G. S. 1996. New Product Team Learning: Developing and Profiting from Your Knowledge Capital. *California Management Review,* 40(4): 74-93..

Markides, C. 2006. Disruptive Innovation: In Need of Better Theory. *Journal of Product Innovation Management,* 23: 19-25.

Mascitelli, R. 2000. From Experience: Harnessing Tacit Knowledge to Achieve Breakthrough Innovation. *Journal of Product and Innovation Management,* 17: 179-193.

McGraw, K. O., & Wong, S. P 1996. Forming Inferences about some Intraclass Correlation Coefficients. *Psychological Methods,* 1: 651–655.

Mednick, S. A. 1962. The Associative Basis of the Creative Process. *Psychological Review,* 69(3): 230–232.

Meyer, A. D. 1982. Adapting to Environmental Jolts. *Administrative Science Quarterly,* 27: 515-537.

Miner, A. S., Bassoff, P., & Moorman, C. 2001. Organizational Improvisation and Learning: A Field Study. *Administrative Science Quarterly,* 46(2): 304-337.

Nijssen, E. J., Hillebrand, B, de Jong, J. P. J., & Kemp, R. G. M. 2012. Strategic Value Assessment and Explorative Learning Opportunities with Customers. *Journal of Product Innovation Management,* 29(S1): 91-102.

Nooteboom, B. 2000. *Learning and Innovation in Organizations and Economies.* Oxford University Press.

Nooteboom, B. 2009. *A Cognitive Theory of the Firm: Learning, Governance and Dynamic Capabilities.* Edward Elgar Publishing Ltd.

Paap, J., & Katz, R. 2004. Anticipating Disruptive Innovation. *Research Technology Management,* 47(5): 13-22.

Park, W. C., Mothersbaugh, D. L., & Feick, L. 1994. Consumer Knowledge Assessment. *Journal of Consumer Research,* 21(1): 71–82.

Paulus, P. B., & Nijstad, B. A. 2003. Group Creativity. In P. B. Paulus & B. A. Nijstad (Eds.), *Group Creativity:* 3-11. Oxford University Press.

Paulus, P. B., & Yang, H. C. 2000. Idea Generation in Groups: A Basis for Creativity in Organizations. *Organizational Behavior and Human Decision Processes,* 82: 76-87.

Piaget, J. 1956. *The Child and Reality: Problems of Genetic Psychology.* New York: Grossman.

Polanyi, M. 1966. *The Tacit Dimension.* Gloucester, MA: Routledge & Kegan Paul.

Quinn, J. B. 1985. Managing Innovation: Controlled Chaos. *Harvard Business Review,* 63: 73–84.

Reid, S. E., & Brentani, U. 2004. The Fuzzy Front End of New Product Development for Discontinuous Innovations: A Theoretical Model. *Journal of Product Innovation Management,* 21: 170-184.

Rice, M. P., O'Connor, G. C., Peters, L. S., & Morone, J. G. 1998. Managing Discontinuous Innovation. *Research Technology Management,* 41(3): 52–8.

Rogers, E. 2003. *The Diffusion of Innovations (5th ed.).* New York: Free Press.

Rosenberg, M. 1956. Cognitive Structure and Attitudinal Affect. *Journal of Abnormal and Social Psychology,* 53: 367-72.

Rosenthal, R., Rosnow, R., & Rubin, D. 2000. *Contrasts and Effect Sizes in Behavioral Research.* Cambridge University Press.

Rubenson, D. L., & Runco, M. A. 1995. The Psychoeconomic View of Creative Work in Groups and Organizations. *Creativity and Innovation Management,* 4: 232-241.

Runco, M. A., & Sakamoto, S. O. 1999. Experimental Studies of Creativity. In R. J. Sternberg (Ed.), *Handbook of Creativity:* 62-92. New York: Cambridge University Press.

Sawyer, R. K. 2005. *Social Emergence.* Cambridge University Press, Cambridge.

Shaw, B. S. 1985. The Role of the Interaction between the User and the Manufacturer in Medical Equipment Innovation. *R&D Management,* 15: 283-292.

Shrout, P. E., & Fleiss, J. L. 1979. Intraclass Correlations: Users in Assessing Rater Reliability. *Psychological Bulletin,* 86: 420–428.

Slater, S. F., & Mohr, J. J. 2006. Successful Development and Commercialization of Technological Innovation: Insights Based on Strategy Type. *Journal of Product Innovation Management,* 23: 26-33.

Slater, S. F., & Narver, J. C. 1998. Customer-led and Market Oriented: Let's not Confuse the Two. *Strategic Management Journal,* 19(10): 1001–1006.

Stasser, G. 1992. Pooling of Unshared Information During Group Discussions. In S. Worchel, W. Wood, & J.A. Simpson (Eds.), *Group process and productivity:* 48-67. Newbury Park, CA: Sage.

Stasser, G., & Titus, W. 1985. Pooling of Unshared Information in Group Decision Making: Biased Information Sampling During Discussion. *Journal of Personality and Social Psychology,* 48: 1467-1478.

Sternberg, R. J. 1998. Cognitive Mechanisms in Human Creativity: Is Variation Blind or Sighted? *Journal of Creative Behavior,* 32: 159-176.

Sternberg, R. J., Kaufman, J. C., & Pretz, J. E. 2003. A Propulsion Model of Creative Leadership. Leadership Quart. 14: 455–473.

Sternberg, R. J., & Lubart, T. I. 1999. The Concept of Creativity: Prospects and Paradigms. In R.J. Sternberg (Ed.), *Handbook of Creativity:* 3-15. New York: Cambridge University Press.

Tabachnick, B. G., & Fidell, L. S. 2007. *Using Multivariate Statistics.* Boston.

Taylor, W. G. K. 1989. The Kirton Adaption - Innovation Inventory: A Re-Examination of the Factor Structure. *Journal of Organizational Behavior,* 10(4): 297-307.

Ulwick, A. W. 2002. Turn Customer Input into Innovation. *Harvard Business Review:* 91-97.

van der Panne, G., van Beers, C. & Kleinknecht, A. 2003. Success and Failure in Innovation: a Literature Review. *International Journal of Innovation Management,* 7: 309–338.

Veryzer, R. W. 1998. Key Factors Affecting Customer Evaluation of Discontinuous New Products. *Journal of Product Innovation Management,* 15(2):136–50.

Visser, F. S., van der Lugt, R. & Stappers, P. J. 2007. Sharing User Experiences in the Product Innovation Process: Participatory Design Needs Participatory Communication. *Creativity and Innovation Management,* 16: 35–45.

von Hippel, E. 1986. Lead Users: a Source of Novel Product Concepts. *Management Science,* 32: 791–805.

von Hippel, E. 1994. Sticky Information and the Locus of Problem-solving: Implications for Innovation. *Management Science,* 40: 429-439.

Voss, J. F., & Post, T. A. 1988. On the Solving of Ill-structured Problems. In M. Chi, R. Glaser & M. Farr (Eds.), *The Nature of Expertise:* 261-285. Hillsdale.

Ward, T. B. 1994. Structured Imagination: the Role of Conceptual Structure in Exemplar Generation. *Cognitive Psychology,* 27: 1-40.

Ward, T. B. 1995. What's Old about New Ideas? In S. M. Smith, T. B. Ward & R. A. Finke (Eds.), *The Creative Cognition Approach:* 157-178. Cambridge, MA: MIT Press.

Ward, T. B. 2004. Cognition, Creativity, and Entrepreneurship. *Journal of Business Venturing,* 19: 173–88.

Ward, T. B. 2007. Creative Cognition as a Window on Creativity. *Methods,* 42: 28-37.

Weisberg, R. W. 1999. Creativity and Knowledge: A Challenge to Theories. In R.J. Sternberg (Ed.), *Handbook of Creativity:* 226-250. New York: Cambridge University Press.

Wiley, J. 1998. Expertise as Mental Set: The Effects of Domain Knowledge in Creative Problem-solving. *Memory & Cognition,* Vol. 26 No. 4: 716-30.

Zaichkowsky, J. L. 1985. Measuring the Involvement Construct. *The Journal of Consumer Research,* 12: 341-352.

Zeidner, M., Matthews, G., & Roberts, R.D. 2004. Emotional Intelligence in the Workplace: A Critical Review. Applied Psychology: An international Review, 53(3): 371–399.

Appendix

Table 3.A: Measurement Items and Construct Characteristics

Construct and Items	Loadings	Cronbachs α
User knowledge		**.86**
How high would you rate your expertise in applications on mobile devices in general?	.86	
How high would you rate your expertise in music applications on mobile devices?	.81	
How interested are you in music applications on mobile devices compared to other people?	.60	
How clear an idea do you have of which characteristics are important in providing you with maximum usage satisfaction when using music applications on mobile devices?	.56	
How high would you rate your expertise in the market for music applications on mobile devices?	.70	
Involvement		**.79**
Music applications on mobile devices are…		
important / unimportant[a],	.71	
useless / useful,	.56	
boring / interesting,		
mean a lot to me / mean nothing to me[a],	.76	
worthless / valuable.	.73	
Creativity		**.83**
How hard or how easy do you think it would be for you to describe yourself as a person who…		
often risks doing things differently,	.79	
has original ideas,	.78	
copes with several ideas at the same time,	.77	
proliferates ideas,	.45	
has fresh perspectives on old problems,	.72	
is stimulating,	.64	
will always think of something when stuck,	.43	
can stand out in disagreement against a group,		
would sooner create than improve,	.42	
likes to vary set routines at a moment's notice,	.42	
needs the stimulation of frequent change.		

Table 3.A: Measurement Items and Construct Characteristics (continued)

Construct and Items	Loadings	Cronbachs α
Empathy		**.78**
I know what moves others.	.49	
I get along well with people that I just met.	.42	
I have a keen sense for the feelings of others.	.84	
I know what to say, so others feel good.	.66	
Problem-solving Expertise		**.73**
I often solve creative problems and questions.	.71	
I am skilled in working on creative problem-solving.	.47	

a
 Item is reverse scored

All items were measured on a five-point Likert scale, except for involvement, which was measured using a seven-point semantic differential; italicized items were excluded due to low loading or causation of an increase in Cronbachs Alpha if left out.

Table 3.B: Use Experience as a Classificatory Variable

Construct and Items	Options
Use Experience	
I use applications on mobile devices.	yes / no
I use music applications on mobile devices. (Filter question for the following two)	yes / no
For how many months have you been using music applications on mobile devices?	open question
How often do you use music applications on mobile devices?	less than once a month / once a month/ once a week / everyday / more than once a day

4 The Playful Ingenuity of Potential Users in Collaboration: Enriched Compensation and Improvisation[5]

Martin Hewing

Abstract Identifying shifts in the needs and behavior of social systems and markets depicts promising new market opportunities for creative organizations. Potential users might be adept in reflecting those shifts and irregularities in collaboration due to their outsider perspective. This article is a systematic inquiry into the unique contribution of potential users and its embodiment in communication in co-creation activities. Following the principles of grounded theory, contributions and strategies of potential users are contrasted with those of current users within varying dyadic compositions. Tentative propositions about the merits and limitations of collaboration with potential users and about emergences in varying dyadic compositions are derived. The resulting core category 'enriched compensation and improvisation' shows that potential users' contributions unintentionally pervade the collaboration with social, cultural and emotional values, which can be regarded as problem-finding as opposed to problem-solving creativity. There is also a process dimension immanent in the merit which facilitates the depth of ideas and how much of it can be made explicit to be understood. Their willingness to improvise leads to a series of ad-hoc changes of initial ideas which foster transformation when combined. Organizational guidelines are derived that endorse the use of critique and the emphasis on depth of ideas as opposed to collaborative consent and quantity of ideas in exploration with potential users.

4.1 Introduction

Organizations struggle to reinvent themselves and sustain a competitive edge through the creation of discontinuous innovation (Bond & Houston, 2003; Leonard-Barton, 1992). The discovery of new opportunities involves unexpected leaps of creativity and insight (Mascitelli, 2000). Discovery is therefore conceptually hard to grasp and difficult to implement as a repeatable process. In explorative efforts, organizations therefore often fall back on established ways of doing things and loose the initial goal to explore new markets for more predictable changes (Birkinshaw et al., 2007). Learning deficiency has a stake in this relapse (Stringer, 2000). Especially the high-technology sector, with its

---

[5] This chapter is an extended version of a previously published paper: Merits of Collaboration with Potential and Current Users In Creative Problem-Solving, Martin Hewing, International Journal of Innovation Management, vol. 17(3), © Imperial College Press, June 2013.

dynamic markets, increasing user demands, and rapid technology advancements requires the constant internalization of new knowledge. Those organizations that succeed in these shifts often possess radically different competences to what incumbents before were relying on (Christensen, 1997, 2006). Collaborating with and learning from users in the early phases of the innovation process has been claimed to be a remedy to mitigate the effect of knowledge gaps by academia and management alike (e.g. Mascitelli, 2000; Mills et al., 1983; Vargo & Lusch, 2004; von Hippel 1986). By harnessing users' tacit knowledge, discontinuities in needs in the environment of the organizations and upcoming real life problems can be unearthed. This recognition gave rise to the idea of users as innovators, whose important assets are their subconscious knowledge and experience. Though obviously not every user will become an innovator themselves, many have a vast untapped potential to create value. Especially potential users, i.e. those who are currently not using the products or services of a domain or are hidden in niche segments, are regarded as insightful external partners due to their impartiality with the respective domain (e.g. Christensen, 2006; Danneels, 2002; Day, 1999). However, absent guidelines about how to involve potential users, what to expect from them, and how to make their tacit knowledge explicit, leads to the recourse to either traditional market research methods or a sole focus on current users being a popular choice in organizational practice. The absence of a satisfactory process and content oriented theory for collaboration with users and particularly with potential users leads to misinterpretations of the value of their integration in academia and poor conduct in practice (Greer & Lei, 2012; Kristensson et al., 2008). This relates directly to the lack of organizational knowledge on how to turn away from the dominant designs mentioned above. A better understanding of the unique merits of potential user collaboration will foster confidence in organizations and thus encourage them to extend their market competences to the edges of their own domain. Collaborative settings thereby have become the center of innovative activities, not only since face-to-face interaction enhances tacit-to-tacit exchange (Mascitelli, 2000; Steyaerd & Bouwen, 2004). In user research, however, the processes for co-constructing, disseminating and utilizing of knowledge in collaboration are poorly understood (Bogers et al., 2010; Greer & Lei, 2012; Kristensson et al., 2008). The Marketing Science Institute (MSI) therefore declared advanced studies of social and cultural user behavior in collaboration a top research priority for 2012 (Bharadwaj et al., 2012). Each user is an inimitable individual, who follows personal agendas, has unique contextual needs and life experiences. This article addresses the following research questions: How do potential users influence collaborative problem-solving as opposed to current users? How are the merits of working with potential users embodied in communication? What emerges from their contributions and what are crucial antecedents with beneficial influence?

Through creative idea discussions between potential and current users, I set out to develop a process and content framework for collaboration with potential users which is derived following the principles of interpretative grounded theory method (Glaser & Strauss, 1967; Steyaerd & Bouwen, 2004; Strauss & Corbin, 1996). The collaborative situation reveals the differences and dynamics between perspectives on a creative problem between their members. As part of a larger research project, this study focuses on the behavior of the individual in a collaborative context, how this behavior affects the creative process and how interaction and the individual itself are affected by what emerges. At stake here is not the creative output as such (e.g. ideas), but the creative process leading to the output of ideas. However, the study also investigates how the process relates to the output by means of triangulation. A parallel study on the creative output showed that individual ideas contributed by a potential user who has collaborated with a current user previously (heterogeneous dyad) showed higher market-discontinuity compared to ideas stemming from current users or to ideas from potential or current users who collaborated in homogeneous dyads. The ideas of the potential users from the heterogeneous dyads, however, did not show considerable differences in terms of the requirement for new technologies. An in-depth inquiry into the micro-genetic collaborative processes which led to this performance can therefore inform organizations and thus help them to make better choices in uncertain explorative activities.

The results reported in this article confirm established theories in the context of collaboration with potential users mainly originating in cognitive psychology, but they also challenge conventional theories and common guidelines in organizational exploration. Indeed, traditionally, research explains the merits of collaboration with inexperienced people by their flexibility of thought (e.g. Frensch & Sternberg, 1989; Koestler, 1964; Rubenson & Runco, 1995; Wiley, 1998). Their individual cognition is not restricted by prior experiences within a domain and can contribute fresh takes on task-inherent problems. Flexibility is regarded as one component of creativity which is often measured in terms of the fluency of divergent ideas (e.g. Guilford, 1956; Mumford, 2003). Creation is thereby regarded as an output coming from the mind of an individual (Smith et al., 2006). Literature on user-involvement has largely adopted this stance in the converse argument that focusing on one's current users only triggers more of the same (e.g. Christensen, 1997, 2006; Chandy & Tellis, 1998; Danneels, 2002; van der Panne et al., 2003; Ulwick, 2002). Conceptually, it is however not clear what exactly constitutes the merits of working with potential users. In the present study, I found the merit of collaboration with potential users to emerge from the collateral pervasion of collaboration with social and cultural predispositions. Potential users' contributions have a strong relation to situations, life habits and emotions as opposed to domain-related artifacts. This can be interpreted as a deliberate search for problems that are meaningful in the context of their lives, before thinking in terms of problem solutions. Especially in collaboration with current users these problems are transformed into tangible ideas through cycles of critique and improvisation. Thus, the benefit of collaboration with potential users is also determined by a procedural trait emerging from a collaborative effort. It is not, in this context, suitable to measure performance in terms of the amount of divergent ideas given rise to. For it is not about the statistical probability of getting one good idea out of many. It is rather about empathy for the unique qualities of the situations and life habits embodied in a discussion that can be extracted by means of a thorough debate between participants. Performance is therefore to be sought in the realm of problem-finding creativity, and in an evaluation of the depth of an idea, its discontinuity in ordinary experiences and the extent to which the idea and the underlying human need have been made explicit. These findings strongly relate to the upcoming, albeit still rather underrepresented, paradigm of sociocultural creativity which regards creativity as emerging from the social sphere (Bruner, 1990; Cole, 1996; Glaveanu, 2011; Sawyer, 2005; Shweder, 1990).

Given the overlap of disciplines relevant to this study, I chose an abductive and interdisciplinary approach. On the one hand, new concepts unique to the research context of collaborative and creative problem-solving with potential users are elaborated, and propositions from the literature are scrutinized in an iterative manner on the other.

4.2 The Potential User

It is necessary to provide some brief clarification of the term potential user. Non-users (e.g. Christensen, 2006; Christensen & Bower, 1996; Danneels, 2003), potential users (e.g. Bonner & Walker, 2004; Fuchs & Schreier, 2011; Shaw, 1985), new users (e.g. Arnold et al., 2011; Danneels, 2002; Henderson, 2006), emerging users (e.g. Day, 1999; Govindarajan et al., 2011; Slater & Mohr, 2006), prospective users (Anthony, 2009; Danneels, 2003) or future users (Bond & Houston, 2003; Chandy & Tellis, 1998) are all terms used in the literature on innovation and technology management to refer to users who are detached from the domain of an organization. Often these terms are used interchangeably. What all these various terms have in common is that the individuals referred to do not currently use a product or service within a particular domain or reside in an unaddressed niche. Yet they do have a need or interest in the domain (Govindarajan & Kopalle, 2006; Christensen, 2006; Hang et al., 2010). Similarly to Hauser et al. (2006), these two traits define the potential user and set him or her apart from the vast majority of non-users. An early integration of potential users into the innovation process becomes part of a deliberate scan for environmental changes that renews the organizational innovation competence (Danneels, 2002; Faems et al., 2005; Meyer, 1982). However, empirical studies on collaboration with potential or generally more distant users are scarce and what constitutes their particular merit or the precise nature of their contribution has not been sufficiently defined. Previous knowledge effects on cognition has raised the question of whether the merit of active collaboration with potential users mainly arises from an absence of concrete mental models or whether potential users actually share common tendencies in approaching creative tasks. These common tendencies, were they to be established by research, would fill the idea of flexibility with concepts unique to the context of potential user collaboration. Furthermore, a prior study showed that the ideas contributed by current users who have collaborated with potential users before did not perform better than the ones drawn from homogeneous dyads. Hence, heterogeneity in user collaboration does not appear to be a sufficient condition for above average performance in explorative tasks. As to the managerial importance of a well-steered exploration, a better understanding of the processes involved in acquiring information from users, and of the intertwinement of individuals and context, is crucial.

4.3 Method

In this study, I sought to understand whether contributions and strategies to creative problem-solving between potential and current users are different and how these differences may lead to the emergence of discontinuities from dominant designs. The research focuses on the behavior of individuals, collectives and the inherent collaborative strategies. Collaborative processes are at the core of the innovative activities (Leonard & Sensiper, 1998; Reid & Bentrani, 2004; Shaw, 1985; Sonnenburg, 2004). So, in vivo observation of collaborations becomes advantageous over more participative methods (Steyaerd & Bouwen, 2004). I conducted idea discussions, in which potential and current users were asked to collaborate with each other in order to come up with new ideas within a given domain. Unlike in group interviews, discussions are not conducted to obtain opinions on pre-assembled problems. Observing discussions is rather conceived as a method to probe into implicit information and behavior of participants and collectives. Additionally, collaborative idea discussions with users are common in organizational practice and can be regarded as observation in the field (Mills et al., 1983; Prahalad & Ramaswamy, 2004; Vargo & Lusch, 2004). The initiation of the task is followed by a communication process which is steered by the users. It is they who decide what to contribute and which dynamic the process can take by its rhythm, the alternation between members or its depth (Steyaerd & Bouwen, 2004). The role of the researcher thereby shifts to a facilitator, extracting information through observation. Observation of interaction in small groups is frequently referred to as the ethnography of communication (Hymes, 1974; Saville-Troike, 2003). This method gives a comprehensive view of creativity in context and uncovers the interaction of perspectives and their natural evolution (Glaveanu, 2011). The idea discussions were complemented by a questionnaire containing two open ended questions, which each participant had to complete after the discussions, regarding the origin of the ideas and the collaboration. In total 68 people were invited to take part. In order to mitigate the effects of the social influences in groups and to be able to analyze collective action more precisely, the discussions were conducted in dyads (Rubenson & Runco, 1995). Three dyadic compositions were possible: (1) a potential user discussing with a potential user, (2) a potential user discussing with a current user and (3) a current user discussing with a current user. The idea discussions were performed in the domain of music applications on mobile devices (e.g. mobile phones, tablets). Although there are many non-users within this domain[6], a large share has some affiliation with music as such, and can thus be defined as potential users.

In order to ensure participation of both potential and current users, a screening questionnaire was developed. Following a purposive sampling approach (Given, 2008; Miles & Huberman, 1994), the questionnaire contained domain-dependent and domain-independent constructs. It was sent to university networks and to the Innovation Forum of the Telekom Innovation Laboratories. There was a participation incentive of ten Euro. Respondents of the questionnaire were classified as either current users, potential users, or as not suitable for the idea discussions. The construct *use experience* (Alba & Hutchinson, 1987) was the main classificatory variable, asking whether respondents used music applications and applications in general, how many music applications they use and how often they use it. In case a respondent indicated to use one or more music applications on a weekly basis, he or she was classified as a current user. Respondents who did not use any applications on mobile devices, but indicated a positive attitude towards music applications were classified as potential users. People's attitudes towards music applications were inquired into through the *involvement* construct (Zaichkowsky, 1985), which encompasses personal interests, values, and needs. 34 participants from each class, i.e. potential and current users, were randomly selected and assigned to dyads.

---

[6] The experiment was conducted before the release of Apple's iPad 2, which resulted in an increase in available music applications for mobile devices of all sorts.

4.3.1 Setting and Data Collection

In total 34 idea discussions were conducted within one week, from which nine consisted of two potential users (inexperienced homogeneous dyad), another nine consisted of two current users (experienced homogeneous dyad) and 16 consisted of one potential and one current user (heterogeneous dyad). Overall, 18 potential and 18 current users participated in a homogeneous dyad, while 16 potential and 16 current users participated in a heterogeneous dyad. At the time of the study, the average age of the participants was 26 and most of them were students (48), whilst the others were employees (11), self-employed (5), unemployed (1) or in miscellaneous occupations (3).

The discussion went through the following successive steps. In the first step researcher and participants introduced themselves to each other. Secondly, the participants were asked to do two so-called ice-breaker tasks. The first task increased the familiarity between the participants and between the researcher and the participants. Familiarity through trust and cohesiveness is conducive for the outcome of creativity tasks (Edmondson, 1999). The second task was a collaborative creativity task only performed by the two participants in a dyad, in order to make sure they felt comfortable articulating their thoughts. This step was followed by the main creative problem-solving task – the idea discussion on new music applications on mobile devices. The task was read and handed out to the participants. In case they forgot some of the wording or components of the task while discussing, they could read it again form the hand-out. It was explained to the participants that the novelty of an idea should be interpreted by what appears new to them and that there are no right or wrong ideas. The task was designed to provide creative space and stimulate divergent thinking (Guilford, 1968; Runco & Sakamoto, 1999). The participants were further asked to articulate and verbalize their thoughts to their discussion partner. The sharing of ideas and thoughts through communication is an important component of collaborative and creative problem-solving (Paulus & Yang, 2000). Moreover, intensive verbalization of their thoughts results in rich data and potential hunches to origins and germs of contributions, strategies and ideas. This step lasted 20 minutes. The researcher was in the room, but withdrawn from the space of the two participants, taking notes from observation. It was important that the participants should not feel guided and also not seek for response (e.g. reassurance). Afterwards, every participant went through an unstructured questionnaire, stating how he or she came to grips with the ideas in the discussion and how he or she felt about the collaboration. Without priming the participant, the open questions give clues if the collaborative engagement changed his or her knowledge and/or perspective. The idea discussions concluded with a short debriefing of the participants.

The discussions were tape-recorded and later transcribed, so were the researcher's observation notes and the responses to the paper-based ex-post questionnaire.

4.3.2 Data Analysis

The whole analysis process was conducted applying the principles of interpretative grounded theory analysis (Glaser & Strauss, 1967; Steyaerd & Bouwen, 2004), the genuinely purpose of which is to make formerly tacit processes explicit (Suddaby, 2006). To get a basic understanding of the individual and collective behavior presented in the study, I read the idea discussions applying open coding within the scientific and qualitative data analysis software Atlas.ti6. Each code was marked according to whether it was uttered by a potential or current user and whether it was someone from a homogeneous or heterogeneous dyad. Simultaneously to the development of the coding scheme, I used research diaries to record notes of striking insights which were transferred into memos within the software. Memos were also used to specify each code and summarize each idea discussion and the new insights it brought. Throughout the open coding, patterns and relationships among codes and code categories were identified and also reflected in memos.

Coding. In further detail, a number of steps were carried out to immerse into the data. Firstly, I drew constant comparisons between the utterances of the participants without paying attention to the dyadic composition. This gave rise to differences and similarities in contributions and led to strategies between potential and current users on a more general account. In a second step, constant comparison was applied to examine how the creative collaboration was influenced by what emerged from its communication and from the dyadic composition. Although differences in dyads were examined, the focus remained on the social process, not on a unit such as the dyad (Glaser, 1978). Tentative propositions were suggested to relate the observed content and processes to underlying assumptions. In a further analysis, these tentative propositions were sought to be disconfirmed and refined by incoming data and tested against already coded data. If evidence appeared to disprove a tentative proposition, it was modified so that it would fit the new insights, resulting in more elaborated propositions. This resulted in quite frequent modifications since I soon became aware that the process was structured by the perceived social context of its members and therefore took multiple and unpredictable forms which needed to be reflected in the proposition. Furthermore, the emerging codes and interpretations were constantly compared to the participants' responses from the ex-post questionnaire to exclude systematic differences between them. Some exemplary statements from the questionnaire are depicted in the findings included below. Finally, a last step involved integrating the emerging categories reflected in the tentative proposition into a unified basic social process framework adapting the paradigm model by Strauss and Corbin (1996). The process framework was considered complete once it could explain each transcript with respect to the potential user.

Exploration and literature. Throughout the analysis and comparison process, I moved iteratively between the extracted data and relevant literature, in order to align and substantiate the coding scheme and develop conceptual dimensions and categories (Gephart, 2004). In particular, creativity theories from the sociocognitive and sociocultural psychology of creativity are relevant to this research. To give an example, the path of least resistance model claims that established cognitive categories and schemata relating to the task domain of a creative problem are used as a starting point of creative processes to reduce cognitive effort (Ward, 1995, 2007). I found evidence for the *path of least resistance* model in the current users' behavior. Indeed, it was interesting to observe whether the value of potential users only arises as a result of an absence of accessible categories and schemata or whether a behavioral equivalence emerges. It became apparent that sociocognitive theories seemed to apply particularly well to the opening phase of the idea discussions; while as time elapsed, sociocultural views seemed to do a better fit on what emerged dynamically. Valuable contributions by individuals in the idea discussions appeared to be woven deeply into the social sphere and contexts. With the idea more prominently in mind, analysis of the collaborative process focused on two main themes. On the one hand, I focused on the personal objectives the participants brought along and how they conflicted with or rather benefited the task-objectives. On the other hand, I examined how collaborative attention and consent were established and how the interactions and the individuals were influenced by what emerged from the collaborative action and context. The use of activities and events as a unit of analysis became particularly suitable with regard to the latter, as it allowed the simultaneous study of the individual and the social context (Rogoff, 1995). To understand the shape of personal objectives and the direction of collaborative creativity, special focus was hence directed towards unique idea episodes[7] within an idea discussion.

Reflexivity and credibility. To ensure reflexivity and credibility of the category coding, interpretations and cornerstones of the emerging framework, a variety of techniques and quality checks were applied. For one, I was supported by a researcher who coded four idea discussions of each dyadic composition independently. Differences in the underlying interpretation of the codes were discussed line-by-line until we came to an agreement. In three cases the utterances within the transcripts were ambiguous. A member check was therefore performed in order to validate the accuracy of the transcripts. As a further quality check of analytical decisions, I frequently presented the main interpretations along with the plain transcripts to fellow researchers in the field of innovation management and psychology in monthly research meetings. In addition, informal one-to-one discussions about emerging categories were held on a weekly basis with colleagues who supported me in conducting the idea discussions and who are professional practitioners within the field of user involvement in innovation. The intent was to resolve differences in perspectives pertaining to underlying insights from the original data through debate. Insights from my professional background in ethnographic activities and creative workshops with users and in the music software industry helped to reflect on memos and interpretations. Additionally, the findings of this study will be related to the results of the parallel study on the creative output of potential users. Thus, I worked with the paradigms of mixed method research (Johnson et al., 2007; Teddlie & Tashakkori, 2011) and triangulation (Denzin, 1989) which, particularly in complex research topics such as creativity, helps to draw a comprehensive picture.

---

[7] I refer to parts of the discussion in which the focus is on the elaboration of an idea or stimulus as idea episode.

4.4 Findings

In the interests of comprehensibilty, the findings are presented here in three different grains of focus[8] (Rogoff, 1995). First, I present the differences in contributions between the two user typologies, potential and current users, and how these contributions and strategies influence the collaborative processes. These differences have been found to originate independently of the dyadic context. This is the individual focus. Secondly, I document how the contributions influenced the interaction and how individual behavior and interaction is influenced by what emerges from the collaboration. This is the collaborative focus. For both, the individual and the collaborative focus, tentative propositions are derived emphasizing the idiosyncrasy of collaboration with potential users. The last step finally integrates the findings of the two grains into a basic social process framework of collaboration with potential users which considers the mutual constitution between the individual and collaborative action.

4.4.1 Individual Focus

Throughout the idea discussions, some contributions by potential users seemed to mark an opposite pole to previously documented utterances of a current user, or vise-versa. The fact that they are related but still represent different characteristics made the differences particularly visible and striking. Contributions that relate to each other were subsumed under a comprehensive and neutral category. In the following, these categories are presented and the particular characteristics they took with potential and current users are emphasized. Furthermore, the categories are roughly presented in order of appearance within the creative process.

Contextual Framing. Potential users had a strong urge to be aware of the dyadic context they were in. They actively sought to establish whether their discussion partner was knowing or as equally unknowing as they were within the domain. This behavior constructed the social context as it provided clues about the other person, their motivations and their relationship to the domain. Different forms of the following question were raised frequently by potential users:

---

[8] I do not refer to these as clearly distinctive levels of analysis since they appear to be mutually connected.

> I do not know much about music applications, just from some friends occasionally. Do you use them? (Potential user 54)

In the collaborative focus, contextual framing is of major importance, as it influences how the dyadic partners approach the creative task and how they interact with each other.

Simplification. At the beginning of the idea discussions, I frequently observed a type of behavior which I categorized as *simplification*. I found that potential and current users sought to simplify the task in different ways. Current users depicted existing music applications as a representative entity to start with. These representatives were used as the point of reference to deviate from and to find improvements or even new ideas for. This tendency can be related to a procedural simplification strategy and to the model of path of least resistance (Ward, 1995). People tend to reduce cognitive resources by drawing on accessible entity exemplars. They also repeated the approach throughout discussions for new idea episodes. A current user categorized his applications on his smart phone to give directions for the creative endeavor as follows:

> Let me have a look on my [name of smart phone]. We were already talking about music players, music creation, [...], what about different options for skins to display music files intuitively? (Current user 43)

The stimuli and ideas derived from such an approach are heavily structured by familiar entities and are thereby likely to contain components of them (Ward, 1995). This behavior is detrimental to the creation of something which strongly diverges from existing knowledge and is well documented in sociocognitive theories of the creative process (Weisberg, 1999). Potential users rather simplified the task itself. Essential parts of the task were omitted. Potential users focused on one or two words inherent in the task and started retrieving stimuli and ideas around it. Given the task of the idea discussion, the stimuli could relate to music, to applications or to mobile devices. This is a declarative simplification strategy. Entities of the problem are discarded to reduce complexity. The potential solution space becomes broader, but it is also more difficult to boil down to a valuable and specific idea. A potential user abstracted her approach as follows:

> To me, music is an art form and expression of creation. I am thinking about ways to create art spontaneously on the go. (Potential user 29)

This participant expressed her individual perception of cultural values that related to the task-domain. The context of being confronted with an ill-defined problem in this case allowed for the concealment and abstraction of task-inherent components. The notion of the value of abstraction (Ward, 1995) - as a converse argument to the path of least resistance model - relates to this behavior and will be detailed in the discussion below. Simplification as a label is used as a comprehensive category encompassing both the codes of procedural simplification and declarative simplification. The tentative proposition I concluded from this unequal behavior between potential and current users is as follows:

> ***Proposition 1.*** *Current users tend to narrow the search field by a procedure of deviation from what already exists, while potential users generalize the problem and find themselves within a broader solution space. The former behavior leads the collaboration to more actionable insights, while the latter tends to lead to more abstract ideas based on individual perceptions and views.*

Critical Incident. Recently experienced events were a major source of information for processing the creative task. Idea episodes often started with such incidents which were recalled and shared by the participants. Current users tended to draw on events they currently experienced with applications, such as flaws or common usage routines. This made the critical incident object-related. The creative task triggered vivid pictures of recent interaction between the person and the application. Indeed, although drawn from a recalled situation, the idea that emerged from the past requirement was directed towards an existing entity. One current user, for instance, started the discussion by describing a problem he had experienced:

> Recently, I did a nice blend of [name of a song] with [name of different song] within [name of the application] and would like to upload it to platforms like [name of application] to get some feedback. These applications are not synchronized or connected. (Current user 58)

Potential users seemed to use a similar creative impulse, which was rooted in recently experienced emotions emerging from a particular situation. The emotions caused the related situation to be vividly recalled. Potential users recalled experiences such as going to a concert, selecting music for a party or trying to sing along to the radio. This made their critical incident context/emotion-related. This category therefore relates to the previously presented category simplification, but is unique in the sense that it triggers vivid associations rather than an unconscious activity. One potential user remembered a recent event in the following words:

> Last week I was sitting in the bus and saw someone nodding his head heavily to the music on his headphones. I got excited and curious and would have loved to have been able to listen to his music. [...] an application that could connect me to his player would be a good and communicative idea. (Potential user 45)

The root of the proposed idea lies in an emotion emerging from a particular situation. As potential users could not draw on negative experience with applications to improve, they tended to spend more time finding a meaningful problem to solve which was derived from life experiences and situations. Current users tended to start thinking in terms of solutions more quickly. I derive the following tentative proposition from this observation:

Proposition 2. While current users' source of stimuli for the creative process tends to be object-related, potential users rely on context-related stimuli and emotions to derive needs and wishes. The former behavior is more likely to lead the collaboration to more obvious problems, while the latter has the potential to formulate a new problem that is rooted in new contexts.

Personal Transfer. As already stated in both preceding categories, current users were more familiar with the particulars of the domain and therefore drew information from their experience (intradomain). In contrast, potential users faced an experience gap. As a result they tended to exploit knowledge from more distant sources (interdomain), such as knowledge from adjacent domains, in which they had experience, interest or even felt passionate about. The distant knowledge was thus related to the task-domain and gave rise to unconventional but also unpredictable combinatorial plays. The transfer of knowledge seemed to be bi-directional - knowledge from other domains was linked to the task domain and stimuli from the task domain were linked to knowledge in other domains. One potential user, for instance, exported the stimuli from the task to the domain of tourism:

> I have been working as a tourist guide and regional music might be something to integrate in guided tours in cities, you know, in these headphones you can rent. When somebody strolls through [name of a borough], she/he would get the music of artists from

this particular part of town or information about which historical music events occurred at that place. (Potential user 112)

What might sound like a fairly continuous improvement from the perspective of the domain of tourism is in fact a rather unconventional usage for the domain of music. The selection of the outside domain was done according to individual preferences and sense of belonging. The potential user seemed to compensate for their experience gap by virtually transferring the task to a domain in which he or she felt more comfortable. The following tentative proposition can therefore be derived from this observation:

Proposition 3. While current users use information from within the domain, potential users tend to compensate their knowledge gap by combining the stimulus inherent in the task-domain with other domains with which they are more familiar. The former behavior ensures established knowledge prevails in collaboration, while the later might result in emergent knowledge and unusual but also unpredictable combinations.

Anticipation. Compensating behaviors that mitigated the knowledge gap within the domain, however, did not always assume a favorable shape. Potential users frequently articulated mere anticipations. They constructed usage scenarios of music applications they had heard of or observed from peers. A potential user initiated a discussion reporting her observation as such:

I know from friends who play in a band that they use this [name of application], but I don't know exactly what it is for. (Potential user 120)

This tended to be knowledge that was not well-structured and vague and mostly seemed to be expressed in situations in which the person felt insecure due to the knowledge gap. Relying on knowledge that is less related to the task therefore seems to require some courage, as it appears to be a less secure base from which to reach a satisfying solution. This category is particularly important in the subsequent collaborative focus and therefore no tentative proposition is derived from it at this point.

Personal Agenda. One alluring contrast relates to objectives of individuals which I coded *personal agenda.* Utterances related to this category most frequently came up when the participants were elaborating in idea episodes. Often, current users did not accept compromise with regard to highly valued performance attributes of existing applications. Imaginative paths that involved significantly changing or completely doing away with an established component valued by the current user would thus usually be dismissed. This behavior is rational, as participants' current usage of music applications is directed towards a purpose and need. Their personal agenda is thus directed towards finding something new and valuable without having to sacrifice the benefits of the status-quo. The agenda however conflicts with the task objective. Potential users, in contrast, were more likely to embrace trade-offs between performance attributes. Furthermore, they paid special attention to the mobile aspect contained in the task. A potential user and a current user from different dyads expressed the following assessments of a similar situation:

> The quality of the recording via a smart phone is most probably low, but it is a good chance to record rough ideas for new songs or make small samples when on the go. (Potential user 73)
>
> A recording function does not make any sense, as the quality of the recording via the microphone is so low. (Current user 114)

The personal agenda category therefore appears to have two levels. The first level relates to the general willingness of a person to sacrifice established performance attributes in favor of others. The second relates to the willingness to temporarily sacrifice established performance attributes in favor of usage in a new context. In the example above, the potential user is willing to compromise quality for the sake of flexibility or mobility. Current users tended to prescind from these (promising) paths of imagination. Potential users, on the other hand, not only embraced such trade-offs, but they also consciously expressed stimuli that would shift established paradigms or foster a redefinition. Some of them originated from barriers the potential users faced within a domain. For instance a potential user expressed his resentment about why people engage by recommendation as follows:

> I absolutely dislike recommendations made by media platforms, as it narrows you down to the same things every time. Let's get rid of that. (Potential user 69)

At times, potential users had personal agendas which were at odds with those of current users as they advocated a replacement of established paradigms by new ones. These shifts can give rise to new ways of solving problems which had been considered solved by people affiliated with the domain and thereby to disruption and substitution of ordinary rules and experiences. I derive the following tentative proposition from the observations on personal agendas:

> **Proposition 4.** *While current users tend to dismiss imaginative paths which alter their current usage habits and contexts, potential users pay deliberate attention to them and embrace redefinitions. The former behavior inhibits change in the collaboration process, while the latter might lead to deviation from common expectations.*

To sum up, the identified contributions and strategies of potential users have two things in common. First, potential users can be seen to enrich the communication with emotional, social and cultural values and views. This happens unintentionally, since the enriched information is rather a collateral benefit of the urge to compensate for an experience gap. Second, potential users tend to find a problem that in the context of their lives proves meaningful to solve, before they start thinking in terms of solutions. This incorporates rethinking solutions to established needs which are considered solved by domain insiders. Table 4.1 depicts further proof quotes from potential and current users.

Table 4.1: Proof Quotes

Category	Potential Users' Code Characteristic & Utterance	Current Users' Code Characteristic & Utterance
Simplification	**Declarative** Mobile phones have microphones, so it is actually a recording device for any kind of signals. You can distort your voice for instance. [...] could that be of any purpose? (Potential user 38) Music is all around nowadays. You actually do not need your own music anymore. That is sad in a way, since one you do not pay attention to something very nice. Let's make an application that changes that. (Potential user 81) For me music is for dancing. I love to go dancing in clubs, but it is somehow hard to find good places that fit my taste of music. (Potential user 108) What are telephones actually capable of? You can do video calls on smartphones, can you? (Potential user 90)	**Procedural** Let us start by thinking about which applications we use and for what reason we use them. When on the go, one should have the same comfort as at home on the computer or in your WiFi. (Current user 107) What applications did we consider: music creation, singing [...]? Information about artists and concerts is something we can still go into. (Current user 5) [Name of application] is good, but I would want to enter parts of the lyrics, so it would recognize the song I have in my head. (Current user 85) Combining music streaming as opposed to downloading it with an application recognizing songs when you record them via the microphone would be helpful. (Current user 10)
	Context/Emotion-related	**Object-related**
Critical Incident	Last week I was at a concert and would have loved to be able to take the music home. The band played their songs really differently than on the records. (Potential user 120) I went to a party that took place in the circle line last weekend. They were going around and around the city and playing music via a sound system. [...] it would be nice to have an application, with which you can connect plenty of phones, so they playback the same song. (Potential user 123) I had some friends over last weekend and as usual we had the problem with the style of music. There are so many different music styles out there now, that it is hard to make everybody happy at a party. But nearly everybody carries their phones with their favorite music; it just needs to be connected and cued somehow. (Potential user 127) I saw [name of movie] for the 1000th time yesterday, and each time I fall in love with the background music, but never really looked up who is the artist behind it. If I could use the camera of a phone to scan the scene, an application could directly download that music to my phone. (Potential user 69)	When I listen to music from my phone, I have the phone in my pocket. Now, when a phone call comes in, I do not know who is calling, so I need to take the phone out. I would like an application that integrates that information into the music. (Current user 118) Every time I plan to buy music, I go on [name of website] to stream the whole song. I would like a store application, where I can stream a complete song once or twice before I download it and pay for it. There is no such thing currently. (Current user 9) I just downloaded an application, which is displaying different instruments that you can play via the touch screen. I would want to add turntables to the selection. It is an instrument after all. (Current user 107) I stopped using equalizers a while ago, since they just don't sound right. So first thing that came to my mind is a better adjustment of equalizer setting within players, maybe even an integration of a whole mixing console. (Current user 1)

Table 4.1: Proof Quotes (continued)

Category	Potential Users' Code Characteristic & Utterance	Current Users' Code Characteristic & Utterance
Personal Transfer	**Interdomain** When I go jogging, I use a device that tells me my pulse and relates it to the phase of my sports plan. I could well imagine that an application could make me aware of my pulse and the phase through music. The rhythm and speed of the song played tells me when I should slow down or speed up. (Potential user 90) [Name of a city square] is for me always associated with the roaring twenties. Maybe locations can trigger typical music of the location on a phone when I pass by these places. (Potential user 112) In my fitness center, I do these courses where the trainer is in front, playing music and we repeat his moves. I am thinking how this could be of interest for music applications. (Potential user 73) I frequently watch videos on [name of video portal], where artists try to teach creation of art like music, or sculptures. Maybe this can be connected to some fun application, like Karaoke, where you are taught how to sing the song properly. (Potential user 91)	**Intradomain** I have plenty of music applications you can use creatively, but most of them have the typical teething problems. Only the simple things work well in this [music application] area. (Current user 114) There is absolutely no application in the market that works like a contact forum for musicians. It would be nice to have something like this in my pocket with me. (Current user 113) Karaoke applications are becoming really fashionable at the moment, but I think they can be improved. (Current user 107) In general there should be more applications addressing musicians. The devices have enough processing power nowadays, but there are not many out there and most of them are pinched. (Current user 65)
Personal Agenda	**Embracing Redefinition** Sometimes it is the limitations of parameters and functions that spark your musical creativity. So I think such functions on a smart phone or tablet could achieve that. (Potential user 122) I would like the idea of renting music, not downloading it for a set price, but for a month or just from your neighbor in the bus. (Potential user 6) Bad internet connection for a mobile application that needs connection can be avoided by providing access to the plenty of unused WiFi networks everybody has in their places. (Potential user 97) I have the feeling that music services and magazines all want to profit from big names of artists. We should make something that particularly focuses on the no-names, the imperfect ones. (Potential user 125)	**Continuity** I only feel safe to make my music downloads within my WiFi network at home. I would never do that on a mobile on the go. What happens if the connection interrupts while you download? (Current user 113) The mobile phone for me is mostly about communication and playing music, I think new functions would disrupt these primary functions. The usability on small screens is too bad for more than that anyway. (Current user 110) Connecting devices to get a louder sound in parks or where ever outside won't work, since the speaker of each phone is just too bad. (Current user 113) I think it is not possible to be creative in the subway for instance. And also, I think you cannot create musical sketches on the mobile phone due to the small screen and storage capacity. I stick to the music library related things. (Current user 118)

4.4.2   Collaborative Focus

I will now report on the idiosyncrasy of the processes within the three different dyadic compositions that emerged, amongst other factors, from the uniqueness of contributions. Here, the contextual framing category plays a central role, as it alerts the participant to the dyadic composition and preconditions of their discussion partner. I will now report on what emerged from the collaborative events and how this influenced the interaction and individual behavior.

Sense-Making. Within the heterogeneous dyad, I observed that participants frequently engaged in a behavior that I refer to as *sense-making*. This refers to people's urge to reassure themselves that they are doing the right thing. This is an act of evaluation and an essential part in human problem-solving. Both classes of users sought assurance or sometimes even tried to justify their utterances. This was particularly striking after a person articulated an idea. Articulating an idea in front of a stranger is an act of exposure. Substantiation of one's utterance is therefore rational. Potential and current users however drew on different sources to establish sense in their articulations. While current users tended to rely on their domain-related knowledge to align their idea with common market understanding, trends or particular white spots within a market, potential users tried to anchor an idea in life habits, further relevant situations and, sometimes, in emotions. The current users' motivation to act as described above is rational behavior stemming from their market affiliation. A relation to common routines that have proven successful in the past attributes usefulness to an idea. Thus, after articulating an idea on a music streaming service, one current user related his idea to current market developments as follows:

> Applications that offer music streaming for a monthly subscription fee are becoming more ubiquitous at the moment. (Current user 43)

In extreme cases, the market affiliation made the current user feel powerless despite the task. Everything that is useful seemed to exist by now. General evaluations of this kind showed how interwoven their experience is with their aspiration. The lack of domain-related knowledge in potential users does not allow them to align their ideas to existing solutions, market trends or white spots in the market. Instead, they frequently engaged in repeating the unique qualities of the situation their idea related to or sought further contexts to which their idea might apply. One potential user for instance introduced a second context to his idea that emerged from a context/emotion-related critical incident:

> Another situation where it becomes useful is driving a car. Imagine, you won't get distracted from driving and the music adapts to your driving style. (Potential user 127)

Furthermore, potential users showed tendencies to emotionalize their ideas:

> The application reacts to your body movement that would be not only experience music but music experiencing you. (Potential user 121)

In other cases potential users tried to fit their ideas into the previously articulated usage contexts of their respective discussion partners. This was part of their attempt to persuade the other of the usefulness of their idea and engage the person emotionally. Since contextual framing unfolded awareness of the dyadic context, a potential user in heterogeneous dyads felt the need to explain him- or herself more thoroughly. Analogies, emotions and further usage contexts stabilize the sense behind the idea and also keep the idea discussion alive. It leads to an enrichment of the initial idea and directs the attention of the collaborating partners to the idea. The current user plays an important part in making the idea of the potential user the focus of attention for an idea episode. Their re-affirmation and qualitative evaluation of the idea is crucial to not only keep the potential user motivated, but also keep the process fruitful in general. Current users articulated their buy-in frequently in varying forms of sentences such as these:

> That is a nice idea and I actually see a demand for such things. (Current user 131)

The described social causation thereby was unique to the context of a heterogeneous dyad. Intense sense-making emerged from the awareness of the contextual frame and the difference in experience levels.

Within the homogeneous dyads of potential users, the participants found out that their discussion partners were equally inexperienced which seemed to be a relief. As a consequence, however, they did not set out to make their idea as thoroughly meaningful in the eye of their discussion partner. Idea episodes seemed to last less long and with less focus before the dyad switched to the next train of thought.

Homogeneous dyads of current users mostly began with ideas derived from current dominant designs. In general, the evaluation of ideas seemed to be of minor importance, since mostly they came to consensus about an idea being useful or not fairly quickly. This consent emerged from their high level of shared information. Consent was also reached with regards to the approach to new ideas. Divergence from dominant designs was often repeated at the beginning of new idea episodes. The process itself appeared coherent and fluent. A current user suggested:

> Let's go on with applications we are using [...]. What haven't we looked at yet? (Current user 107)

The need to explain oneself to the other appeared circumstantial. The homogeneous dyads went through more unique idea episodes, while the heterogeneous dyads in contrast focused on fewer stimuli due to intensive sense-making and comprehension obstacles that required thorough interaction. I derive the following tentative proposition from the observed behavior:

Proposition 5. *Within heterogeneous contexts, potential users spend deliberate time to convey their idea and make it relevant in the eyes of their discussion partner. This internal selling draws collaborative attention to the idea and helps to explicate the tacit information defining the stimuli in more depth. In homogeneous contexts, mutual experience leads to quick consent on predictable entities (current users) and mutual inexperience leads to shallow elaboration of initial stimuli (potential users).*

Assimilation. Within the heterogeneous dyads, contextual framing was frequently followed by a phase in which the differences in the knowledge bases between the participants was mitigated. Contextual framing led to the recognition that the current user's domain knowledge is unique. The potential user would actively claim information on the domain from the current user or the current user would voluntarily share information from his or her insider perspective. Both ways enabled the potential user to become familiar with the domain. In cases whereby the potential user anticipated usage scenarios from vague knowledge - as described in the individual focus - the current user could comment on it and put things right. This made the process more efficient compared to the homogenous dyad of potential users, in which the anticipation often led to corrupted idea episodes. The anticipation was carried through from the scanning of information, to the articulation of an idea and its evaluation. A statement of a potential user from a homogeneous dyad makes this clear:

> It would be nice if you could recognize songs by simply recording something you are listening to currently. [...] though I think this already exists. (Potential user 128)

In this dyadic composition, the discussion partner is not a remedy to mere anticipation. His or her knowledge is at most vague and imprecise, too. The process becomes inefficient and the idea episode comes to a standstill. When the inefficiency of this approach became apparent to the participants, they altered their approach to what I labeled declarative simplification in the precedent paragraph. They omitted task-inherent categories and schemes and thereby broadened the solution space. The following suggestion of a potential user made the alteration explicit:

> Applications do not need to relate to music, there are different ones such as games or weather reports. Maybe we can start from there. (Potential user 108)

Within the heterogeneous dyads, assimilation was not only important to the potential but also to the current users involved. In some cases I observed a process of unlearning and a change of perspectives. These cases became apparent by utterances of current users who stated a positive evaluation of utterances from potential users. A current user evaluated the stimuli of a potential user as such:

> That is a good point. I have not seen it from this perspective before. Let us try to build our idea on this. (Current user 72)

Not only did the current user like the unfamiliar thought, but also identified it as a good path for further elaboration on the idea. This led to a collective cognition, which preceded a fruitful elaboration process. Further evidence for assimilation appeared to crystallize itself in the abundance of comprehension questions current users asked their respective discussion partners and vice versa in heterogeneous dyads. This is an indicator for comprehension problems and the need for thorough understanding from both sides. Contributions by potential users, as presented in the preceding paragraph, helped to mitigate distance between potential and current users. Both participants can easily relate to more abstract or emotional stimuli, such as going to a concert (e.g. critical incident). This directs the collaborative attention away from more specialized approaches focusing on domain-related artifacts towards more emotional and real-life contexts. The following tentative proposition therefore can be derived from this observation:

Proposition 6. *In heterogeneous contexts, learning, unlearning and mutual guidance between members leads to collective cognition with a focus on life habits and emotionally bound starting points for idea elaboration (idea episodes). In homogeneous contexts, the participants come to a more predictable consent (current users) or to an inefficient process leading to abstraction (potential users).*

Deliberate Debate. This is a category that relates exclusively to the heterogeneous dyad. Once a collective cognition was achieved, idea episodes seemed to be structured by the personal agendas of the current users and improvised collaboration. Nevertheless, a current user approved of the idea of a potential user, his or her urge for alignment to dominant designs and more generally his or her personal needs and objectives did not founder. Questions pertaining to the implementation or potential obstacles of an idea were raised. A current user for instance expressed a common legal problem after the potential user articulated an idea:

> You will encounter legal problems when you post mood playlists on the internet, since you are not allowed to share music freely. How could that be solved? (Current user 96)

The critique of the current user appears as a creative challenge to the potential user. In the above example, the potential user responded like this:

> Yes, maybe, [...]. But maybe we do not have to post playlists, but we post song tags. Radio is a free medium and there are so many radio stations out there worldwide that can be listened to via the internet. What if the application searches radio stations worldwide to find whether the song from the playlist is being played somewhere at that very moment and blends over into the radio station if it is actually found? Wouldn't that deal with the legal restraint? (Potential user 121)

The potential user takes the critique as a creative challenge and tries to circumvent the problem. Potential users found comfort in the ill-definition of the problem and started to improvise. Therefore, critique, if constructive, should not by all means be avoided in creative problem-solving in this context. Quite the contrary, it can even be found to fuel the process to arrive at more unusual but useful insights and solutions.

When a current user suggested that a potential user's idea would not work properly in the mobile context due to a deficiency in the stability of mobile internet, the current user suggested to dismiss the idea path altogether. To him it seemed a pointless idea without the same stability of service he was used to from home WiFi. The potential user however quickly altered his idea:

> There are plenty of WiFi hotspots across town, so once you walk past one, your mobile device connects to it and exchanges playlists with your computer at home or your friends'.[...] Or, even better, other users of the application can unlock their home WiFi's and you exchange playlists with them when you pass close to their place. That makes the use of the application even more exciting and social. It is also a fun way to discover new music literally on the go. (Potential user 97)

Here, the current user's personal agenda was perceived as a stimulating challenge by the potential user. Constructive critique of the current user triggered new divergent processes in which ideas were altered according to incoming requirements. The potential user in the example above responded with an alteration of the original idea and could convince the current user to go on with the elaboration. His alteration of the idea also added new and unexpected properties to the idea – to include the social exchange and discovery of new music as opposed to a mere exchange of playlists with one's own computer at home on the go. Each utterance in the communication can encompass an additional idea or a challenge to the utterance from before. In other cases critique or challenges are reframed in order to allow for positive twists. A current user shared his experience and complaints about mobiles phones or tablets in music production:

> The limited screen size makes the use of music production tools on a mobile phone obsolete. These programs have so many parameters. (Current user 110)

The potential user rapidly infused new meaning to the critique:

> Sometimes it is the limitations of parameters and functions that spark your musical creativity. A reduction to important functions on theses mobile devices could well achieve that. (Potential user 122)

The proposed opportunity does not come from a concrete idea, but from changing meaning. Improvisation is, therefore, embodied in (1) circumvention of the critique by altering the idea or (2) redefinition of the critique with novel meaning turning critique into a perceived opportunity. The debate transfers knowledge, member preferences and helps to extract elusive information within a series of single insights. The following tentative proposition is derived from this observation:

Proposition 7. Within heterogeneous contexts, unique information of each individual is shared. Sharing allows the collective cognition to evolve and gives rise to a deliberate debate, in which critique and support co-exist. The debate attributes meaning to the stimulus through a series of alterations and redefinitions of the initial idea in a collaborative effort.

In extreme cases, the friction of personal objectives led to dissonance between the participants that needed to be solved in order to proceed.

In sum, especially the heterogeneous dyads had to establish collective cognition. In homogeneous dyads it existed partly due to the shared experience or inexperience of its members. Once the heterogeneous dyads achieved a collective cognition through learning and unlearning, the uniquely held information led to benefits. The deliberate debate with inherent cycles of critique and circumvention or redefinition substantiates and extracts the unique qualities of the enriched insights to tangible ideas.

Table 4.2 lists original statements from the ex-post questionnaire. The first question was directed towards the origin of the ideas and the answers reflect the categories found in the individual focus. Answers to the second question with regards to the personal experience of the participants within the collaboration reflected categories emerging from the collaborative focus.

Table 4.2: Proof Quotes from Questionnaire

What was the impulse for your ideas?	How did you experience the collaboration with your discussion partner?
Potential Users from Homogeneous Dyads	
Thinking of everyday situations helped a lot to get familiar with the task. (Potential user 20) → *Critical Incident (context/emotion-related)*	It was not an easy task, since we both are kind of rookies to music applications. (Potential user 51) → *Anticipation*
I am not so familiar with applications for music; I tried to come up with crazy things. (Potential user 18) → *Simplification (declarative)*	We ended up thinking of how people can be connected to each other. That's not really a music thing, is it? (Potential user 120) → *Abstraction*
Potential Users from Heterogeneous Dyads	
I tried to generate ideas from my personal interest and emotions when listening to music. (Potential user 11) → *Personal Transfer (interdomain)*	My discussion partner had a different opinion on and interest in music, which somehow inspired me. (Potential user 30) → *Assimilation & Collective Cognition*
I was thinking of what I like about music and used the impulses of my discussion partner who had more experience in the domain. (Potential user 69) → *Simplification (declarative)*	My discussion partner showed me, that all the quick ideas that came to my mind initially already existed. (Potential user 17) → *Assimilation & Collective Cognition*
Current Users from Homogeneous Dyads	
The lack in quality of current applications which constrains the ease of use needs to be abolished. (Current user 89) → *Simplification (procedural)*	Since we had the same interests, it was easy to convey our ideas to each other. (Current user 65) → *Predictable Consent*
I was thinking of my [name of mobile phone] and applications that I currently use. (Current user 9) → *Personal Transfer (intradomain)*	We had somehow the same problems with current applications which made the exchange easy. (Current user 24) → *Predictable Consent*
Current Users from Heterogeneous Dyads	
Everything useful actually does exist already. The only improvement to be made is a more comfortable use. (Current user 84) → *Personal Agenda (continuity)*	The ideas of my partner were surprisingly interesting and convincing. (Current user 88) → *Assimilation & Collective Cognition*
I was thinking of my own usage of music applications and what I am missing within them. (Current user 95) → *Critical Incident (object-related)*	The development and the discussion of ideas with my partner were very helpful and got me to understand a different perspective. (Current user 96) → *Assimilation & Collective Cognition*

4.5 Collaboration with Potential Users as a Basic Social Process

Throughout the analysis an integrative framework outlining the creative collaboration with potential users revolved around their tendency to enrich the communication with social, cultural and emotional values and their composure to improvise. It quickly became apparent that the individual and the collaborative focus are mutually defined and interdependent (Rogoff, 1995). I subsumed the observations of collaborations with potential users under the core category *enriched compensation and improvisation*. The core category illustrates a two-step process which is to be explained in further detail.

Enriched compensation. The propositions of the former analysis relate to the categories simplification, critical incident, personal transfer, personal agenda, sense-making and assimilation and are not only actions to compensate the gap in experience within the task-inherent domain. They also have a strong and eventually unexpected tendency to pervade the collaboration with information that is individual and personal. This relates to individually experienced situations, daily life habits and emotions as opposed to domain-related artifacts. The information however does not become part of the interactional frame until it is evaluated by the other participant in assimilation (Sawyer, 2005). It requires reaffirmation, acceptance and eventually unlearning from the other person. Assimilation as a consequence of contextual framing is thereby crucial as is the process by which the person recognizes the plurality of perspectives and conceives of its own position within it (Jovchelovitch, 2007). The individual is influenced by what emerges from its own prior communication. In heterogeneous dyads, enriched compensation emerges in its strongest form. Sense-making and assimilation behavior are particularly prevalent in this context and not only enrich the utterances, but also direct the attention of the collaboration away from artifact-related information by engaging the participants emotionally. The collaboration thus comes to a collective cognition and immerses into the unique qualities of these constituents in a deliberate debate within idea episodes.

Enriched improvisation. This latter part of the core category emerges from the enriched compensation behavior, which constructed a mutually perceived safety zone allowing for critique and support simultaneously. Personal agendas of the current user are perceived as challenges by the potential user and again fuel a creative process within unique idea episodes. A deliberate debate emerges substantiating the elusive qualities inherent in the enriched information. I called it a debate, since the individuals can take perspectives within their collective cognition. It also stimulates improvisational play, as each utterance can be an additional idea or critique that becomes part of the frame. Improvisation does not allow for an anticipation of the dynamically evolving collective cognition. New information for both members emerges and can influence their subsequent actions (Sawyer, 2005). The process therefore arises from a collaborative effort. Discontinuity of the ordinary experience embedded in the improvisational dialogue comes from (1) circumvention of problems inherent to the idea or (2) a redefinition of the meaning of the critique as depicted exemplarily in Table 4.3. The nature of the problem (ill-definition) further stimulates improvisation. The absence of an idealized end-state encourages the collaboration to improvise. The improvisational play shifts the relevance and the merit of collaboration with potential users away from the ultimate output of the creative process towards the process and communication itself. Within enriched compensation the task becomes finding a meaningful problem to solve, while enriched improvisation concerns substantiation, extraction and formalization of the unique insight.

Communication in the heterogeneous settings is less predictable than in the homogeneous settings. Personal objectives collide and open new paths of imagination. The problem inherent in the task is processed to a new state. Hence, the merits of collaboration with potential users are rooted within these processes and the contents that they bring up. Homogeneous dyads are more predictable in their activity, feature less emergent properties, but go through more unique idea episodes. Dyads of potential users are ineffective by anticipation at first and then more abstract in content, with less focus on thorough exchange. Dyads of current users reach consent easily and take comfort in accessible knowledge, which also leads to repetition of the creative approach throughout unique idea episodes. Their process and outcome is relatively easy to follow and conveys an amplitude of unique idea episodes.

Table 4.3: Example of Cycles of Critique, Circumvention and Redefinition

User	Code	Utterance
Potential	Idea Articulation	[...] An application that lets you exchange the whole music library on your phone when bouncing the phones together could be a fun way to explore music and meet new people.
	Redefinition	It is like lending music and not buying it. [...]
Current	Critique	[...] there is also a legal problem of exchanging the music for free [...] and who would still pay for music, when it is all around you?
Potential	Circumvention	People need to pay a leasing fee or even better...we could add a donation function. People can donate money directly to the artist if they like the music they get.
Current	Critique	If they don't have to I guess nobody will do it? They need to be forced.
Potential	Circumvention	The application could suggest donating a self-determined amount of money to the artist once the song is played more than five times for instance.
	Redefinition	It might even tackle the money problem of the music industry, since people rather donate than buy things. It gives them an altruistic and sophisticated feeling to support art. [...]

It is seldom that one sole condition causes a phenomenon (Strauss & Corbin, 1996). Enriched compensation and improvisation is therefore an abstract label for the process and its varying predictability throughout the observations. The process which this label describes is cyclical rather than linear and features a series of insights and actions that lead to something meaningful. The propositions presented in the findings depict snapshots, which taken on their own might not lead to the desired impact. They go hand in hand and some of them are the consequences of the other. Figure 4.1 integrates the categories that arose from collaboration with potential users into a process framework[9].

---

[9] Note that the process does not enlist the concepts uniquely attributed to the current user, as the purpose of this paper is a better understanding of the potential user. The influence of the current user is considered within the cognitive distance of the discussion partner as a contextual determinant.

Figure 4.1: A Situated Framework for Collaboration with Potential Users

Nature of the Problem (e.g. ill-defined); Collaborative Setting (e.g. face-to-face)

Cognitive Distance	**Enriched Compensation**	Sense-Making	**Enriched Improvisation**
Contextual Framing		Assimilation	

Simplification *(declarative)*	Deliberate Debate
Critical Incident *(emotion-related)*	
Personal Transfer *(interdomain)*	
Personal Agenda *(embracing-redefinition)*	

Anticipation	Abstraction

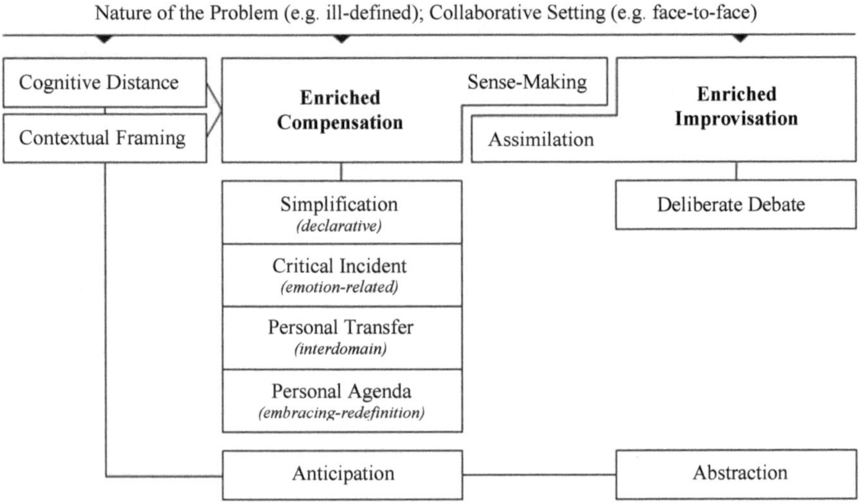

4.6 General Discussion

The findings improve the understanding of the rich and complex insights that arise when collaborating with potential users in creative problem-solving. Although researchers in innovation management have continuously referred to the benefits of collaboration with potential users in exploration (e.g. Chandy & Tellis, 1998; Christensen, 2006; Lau et al., 2010), there has not been a systematic inquiry into the uniqueness of their contributions and strategies. In general, I found the merits of collaboration with potential users to be (1) rooted in the enriched collateral benefits of compensation pervading the collaboration and (2) in emergences from cycles of critique and improvisation. The collaboration leads to a naturally occurring process of finding a meaningful problem before coming to solutions in a series of single insights leading to transformation. The conventional and vague intuition of uncovering unfulfilled needs from (potential) users (e.g. Danneels, 2002; Mascitelli, 2000) is charged with how this uncovering process is embodied within collaborative communication. The merit is thereby partly determined by a procedural trait. Besides the frequently cited theories from sociocognitive psychology, the results show a strong tie to the cultural stance in psychology and creativity research, often referred to as the sociocultural paradigm (Bruner, 1990; Cole, 1996; Glaveanu, 2011; Sawyer, 2005; Shweder, 1990). I consequently directed the discussion of theoretical contributions in particular to the context of creative organizational exploration and its relation to both creativity paradigms, the sociocognitive and the sociocultural. Furthermore, in the vein of mixed method research (Johnson et al., 2007; Teddlie & Tashakkori, 2011) and triangulation (Denzin, 1989), these findings are related to the results of a parallel study that focused on the creative output of collaborations with potential users as explained earlier on.

4.6.1 Theoretical Contributions

The list of research investigating the behavior and even more so the outputs of more or less knowledgeable persons in problem-solving is long and spreads over various disciplines (Weisberg, 1999). The effects of previous knowledge on creativity are undisputed in both the sociocognitive and the sociocultural paradigm. Research on behavior in collaborations as opposed to individuals in creative problem-solving, and on ill-defined as opposed to well-defined problems is already more scarce (Sonnenburg, 2004; Voss & Post 1988). Empirical research that combines the aforementioned attributes to the context of user involvement in innovation is thereby bound to quantitative research, a focus on the creative output and an organizational perspective rather than an individual or small collective one (Arnold et al., 2011; Bonner & Walker, 2004; Chandy & Tellis, 1998; Danneels, 2003; Govindarajan et al., 2011). The in-depth inquiry therefore sharpens the view of user collaboration particularly with potential users in the context of discontinuous new product and service development and provides clues towards a content and process theory from which guidelines for organizational exploration can be derived.

The enriched compensation strongly relate to and partly support theories from the sociocognitive paradigm of creativity within the context of collaboration with potential users. Conceptually, they address the relation and tension between knowledge and creativity and the involved level of cognitive flexibility (e.g. Frensch & Sternberg, 1989; Koestler, 1964; Rubenson & Runco, 1995; Ward, 1995). The process of pervading the collaboration with information that is enriched by social, cultural and emotional values is to a large extent however collateral. It therefore became apparent that the merit is also determined by emergent properties, which cannot be explained solely by the composition of the collaborative members and their mental models. Emergent properties are usually associated with unintended effects of action (Sawyer, 2005; Ward, 2007). Not only did the individuals as social entities influence the dyad, but the social situation influenced and constituted the individual in this context. Especially in heterogeneous contexts, a collective cognition was reached by sense-making and assimilation which directed the collaborative attention towards the intangible aspects of a meaningful problem. It might be argued that assimilation is a form of group convergence studied in sociocognitive psychology (e.g. Larey & Paulus, 1999). Members of groups tend to direct attention to information that its members have in common. Even more so, however, it is a process of learning and unlearning, which is a property of the group, not the individual (Hutchins, 1995; Sawyer, 2005). The deliberate debate and its improvisational dialogue is an emergent property. Each utterance of the improvisation is unforeseen, but influences the collective cognition and the idea in discussion. The merit therefore does not arise from individual minds and their interaction, but from something that emerges from the interaction itself. These emergences subsequently enable the individuals to extract the unique qualities of the stimulus. This relates to a more participatory stance in sociocultural studies by Rogoff (1995) in which:

> "participatory appropriation is the personal process by which, through engagement in an activity, individuals change and handle a later situation in ways prepared by their own participation in the previous situation." (Rogoff, 1995).

The results reflect the intricate relations between individuals and environment, which are basic to the sociocultural stance in psychology (Bruner, 1990; Cole, 1996; Shweder, 1990). Researchers agree that the individual and the group cannot be studied in isolation but only in situated practice. Creativity is therefore situated in the social sphere rather than in individual cognition (Glaveanu, 2011; Sawyer, 2005). Hence, this study is a first attempt to strengthen and benefit from the achievements of the sociocultural view in the context of user collaboration.

Expressed somewhat provocatively, the potential user's natural composure to improvise can be compared to the lightheartedness with which children tackle a creative task. Children start improvising more naturally, while adults tend to plan thoroughly before they perform (Baker-Sennett et al., 1992; Bearison & Dorval, 2001). Children unconsciously accept that not all properties of an output need to be elaborated conceptually before it can be shared, tested and also radically changed to obstacles faced on the go. The deliberate debate and its inherent improvisational dialogue relate to this real-time process. Furthermore, communication and thorough collaboration become crucial to children, since they lack experience and mental models about artifacts (Kratzer & Lettl, 2008). This notion is reflected in the sense-making and assimilation category. Potential users depend on thorough exchange. Improvisational learning theories in organizational research in general and in new product development in particular acknowledge the relation between improvisation and organizational learning (Crossan, 1998; Cunha et al., 1999; Kyriakopoulos, 2011; Moorman & Miner, 1998). The cycles of critique and its (1) circumvention or (2) redefinition in potential user collaboration can be compared to compression strategies (Eisenhardt & Tabritz, 1995) and contemporaneous advice of composition and performance (Moorman & Miner, 1998). A series of small adaptations leads to transformation. The circumvention is an *artifactual production* (Miner et al., 2001), since the idea is altered or an essential component is added. The redefinition is an *interpretative production* infusing new meaning to the critique and opening new opportunities. This finding also relates to Dunbar's (1997) experiment concerning scientific discovery, in which a series of small mutations in concepts produce conceptual change. It partly unravels the mystified conception of a sudden insight arising through cognitive flexibility (Dunbar, 1997). However, by definition improvisation is not good or bad for the output, as it can lead to originality but also to chaos (Vera & Crossan, 2001). The deliberate debate emerges from antecedents that in this context predominately lead to an original output. Positive antecedents are real-time and face-to-face exchange (Vera & Crossan, 2001), the enrichment of the information by finding real-life problems before thinking in terms of solutions as a starting point for elaboration, and the rise of a collective cognition through sense-making and assimilation. The nature of the problem (ill-definition) is an intervening condition that stimulates improvisation. Despite being a rather unpredictable process, improvisation can thus be shown to be desirable in user collaboration when certain dispositions are met in advance.

Furthermore, the establishment of a collective cognition and deliberate debate take time. Hence, heterogeneous dyads went through fewer unique idea episodes than the homogeneous dyads. Sociocognitve studies measure creative outputs amongst others in the fluency of divergent ideas (e.g. Guilford, 1956; Mumford, 2003). Though the contributed ideas of potential users from heterogeneous dyads were most discontinuous to dominant designs, the users are not necessarily able to repeat that specification due to a dearth of unique idea episodes. The merit of collaboration with potential users is therefore not a higher amount of divergent ideas per se, but has to do with the deliberate debate that emerges and specifies the idea. Extraction of rich and useful stimuli and ideas as both parties answer for their views, opinions and needs is the ultimate gain. This finding relates to the basic assumption of socioculturalism, according to which the sharing of perspectives enables the construction of new perspectives and thereby of discontinuities (Glaveanu, 2011) [10]. The salient ideas of the potential users from heterogeneous dyads had their origin in the unintended and elaborate social process that emerged from a collaborative effort with a current user.

The enrichment of the collaboration with social, cultural and emotional values is also reflected in the study of the output of the idea discussions, in which the ideas contributed by potential users from heterogeneous dyads - though rated more original and surprising - did not require considerable technological leaps. The ideas integrate common technology into the environment in a novel fashion and trigger new usage contexts. This trait relates to innovations that leverage technology suggested by Danneels (2002) in the competence based typology of new products and services. From the perspective of the organization, the merit of collaboration with potential users can thus be seen in learning how to delink one's technological competence from an established product and service and relink it to

---

[10] In the sociocognitive view sharing is rather attributed to decoding and encoding that takes place inside the mind of the individual (Paulus & Nijstad, 2003).

new needs and contexts. The category of the analysis coded personal agenda shows the tendency of potential users to rethink solved problems. It supports Christensen's (1997) notion of disruptive innovations. Disruptive innovations introduce a different set of features and performance attributes which first appeal to a niche market due to their (technical) inferiority on the product trait most valued by the established market (Govindarajan & Kopalle, 2006). Potential users give these disruptive paths of imagination deliberate space in the creative collaboration while current users seek discontinuity without sacrificing their status-quo. When asked to specify one idea in more detail, current users specified an idea that was evaluated as more sustainable to dominant designs compared to those specified by potential users. It is insufficient to assume that potential users underwent further development as a result of the idea discussion and that the current users did not (Rogoff, 1995). Current users acted in accordance with their familiar personal agenda and needs. It is an answer for internal consistency which is a cognitive urge postulated in consistency theories (e.g. Festinger, 1957; Rosenberg, 1956). The basic assumption of these theories is the need of the individual for consistency between attitudes and behaviors. An alternative explanation can be found in the nature of knowledge shared and understanding problems. Current users might have had difficulty formalizing verbally the discussed ideas stimulated by the potential users due to the stickiness of information they contain (von Hippel, 1994). Hence, a potential user benefits from the unique information of a current user, while the current user needs more effort to (un-)learn. Nevertheless, their contribution to the process and its content itself is indisputable.

The homogeneous dyads of potential users accentuated the anticipation of dominant designs which resulted in a certain inefficiency of the process at first. The dyad's subsequent turn to a declarative simplification approach contained somewhat promising stimuli which were not accurately extracted and therefore not mirrored within the idea outputs. The value of abstraction notion (Ward, 1995) states that abstraction is supportive for innovative creation. It seems intuitive and corroborated by empirical evidence that someone who starts imagining from an abstract scheme will be more likely to arrive at innovative solutions judged as deviating from common characteristics of the dominant design than someone who starts from more specific entities. The potential users in homogeneous dyads embarked on abstraction, but it did not result in a better performance within the process and its output. The value of abstraction can thus only be captured when the context offers possibilities to pursue a fruitful path that extracts and subsequently formalizes the unique qualities of the abstraction. Valuable insights in abstraction generate new questions (Sawyer, 2006) which need to be tackled in debate. Sense-making, assimilation and the deliberate debate achieved this in heterogeneous dyads. These processes made the enriched content more explicit and deliverable.

Both paradigms, the sociocognitive and the sociocultural, are strongly reflected within the study and the findings. Some stages of the idea discussion marked learned behavior from experience while others only became recognized by the individuals when they went through it (Glaser, 1978). Given the fact that most innovative activities nowadays are carried out in collaboration, I suggest that more emphasis be placed on contextual approaches to creative problem-solving, especially when the goal is to identify promising ways of interacting between organizations and close and distant environments. Though creative leaps are often rather unpredictable in their content (Campbell, 1960; Litchfield, 2008), there seems to be a lot to learn from studying the co-construction processes that lead to their emergence.

4.6.2 Practical Implications

Organizations searching for new market opportunities should be aware of the fact that, just as selecting the best idea is crucial for a successful innovation process, directing attention to and acquiring understanding of the most fruitful imagination path for new opportunities is the homework that needs to be done before. It is not that organizations necessarily lack good ideas, but they need to find ways to process and shape these ideas into promising opportunities (Florén & Frishammar, 2012). From collaboration with potential users, organizations can learn how to make better choices in the front-end process of innovation. Potential users tend to define a problem that is meaningful in the context of their lives, before thinking of solutions. The virtue of this perspective resides in the contribution of emotional and experiential insights and the people's playfulness to improvise. As the former does not relate to physical artifacts, these insights are hard to extract, but they can convey rich and eventually surprising elements of information for the organization. The information deviates from expectations and reflects potential shifts in behavior of societies that might become more attractive to a larger amount of people over time. Intense collaboration and particular focus thereby should be on mutual learning which helps the potential user to align and articulate their tacit information and create a collective cognition by engaging others emotionally.

From a perspective of probability, organizations that facilitate creative collaboration activities have long followed the postulate of going for quantity rather than quality of ideas. The underlying statistical reasoning for this is that the more ideas have been conceived at the end of the day, the higher the probability that they will cover a diverse spectrum and that at least a few will be new and surprising (Goldenberg et al., 1999; Simonton, 1977). The study of the merits of potential user collaboration points in a different direction, however. A shift towards fewer ideas but with more depth and elaboration might well result in a more effectively steered exploration. Collaborations with potential user should be given focus and debate. It is not enough to invite potential users and hope for their flexible cognition to take over.

Furthermore, research on group creativity postulates that one should defer any critique in idea generation (Osborn, 1963). The logic behind this is that critique can intimidate people and as a consequence they stop articulating their thoughts freely. It constricts trains of thought and narrows the solution space. This view, however, ignores two essential qualities that critique, when reported constructively, can contribute. On the one hand it transports knowledge and therefore stimulates mutual learning and unlearning in collaboration. This knowledge can serve as a catalyst to extract tacit information and to achieve useful insights. Outputs are creative when they are new and useful (Amabile, 1983). Spending time on fewer thoughts and giving rise to constructive critique will help to understand the unique qualities of the contexts. Instead of getting more of the same, a relation to new and useful contexts will be highlighted aligning ideas and leveraging technological competences to real-life problems. On the other hand, if perceived as a challenge, constructive critique once again fuels the process of imagination and fosters deviation from the norm. I called this process a deliberate debate which frequently emerged in the heterogeneous dyads I studied and led to a series of single insights which, taken together, stimulated meaningful ideas. In exploration workshops, small teams of users should therefore be given deliberate time to discuss ideas in depth. The role of the key individual of the organization shifts to a facilitator. This person's task is to identify meaning, insight and inspiration within emotionally and situationally bound communication, and to carry out follow up steps, such as member checks. Good creative insight often generates more questions than it answers (Sawyer, 2006), but is valuable as long as these questions are stimulating. Tolerating unpredictability and improvisation are thus indispensable key factors.

Methodologically, the internet nowadays allows for new and exciting ways to collaborate. However, collaborations with potential users yield information with particularly elusive and tacit dimensions, which are based on bodily experiences and emotional involvement (Mascitelli, 2000). Physical co-location and face-to-face collaboration therefore seem favorable.

4.6.3 Limitations and Future Research

The merits of collaboration with potential users are particularly embedded in the emergent processes. Particular emphasis was attributed to the social origin of creation and the contextual frame. Universal laws that hold across all contexts therefore cannot be derived, which is however a part of the finding and a consequence of the sociocultural stance. Especially the context of heterogeneous dyads resulted in rather unpredictable interactional frames, which I summarized in abstract categories in order to fit to all observations. Changes in the setting of this study thus point to promising research paths. The data came from observations of interaction processes, which can be called ethnography of communication (Hymes, 1974; Saville-Troike, 2003). In organizational practice, user involvement does not always require the user to be involved actively. The study and comparison of potential users in the continuum between passive and active participation therefore leaves room for various interesting study designs. Furthermore, the task given to the discussion members was ill-defined. It had the inherent claim to discuss really new products and services as opposed to sustaining the current designs. The ill-definition of the problem gave comfort and rise to improvisational processes, but it also restricts the findings. Collaboration with potential users in more sustainable exploration activities is not covered here. The ill-definition is a component of the situated action framing the idea discussions, so are others, e.g. the domain. This study focused on one broad domain which can take various degrees of complexity. Communication within heterogeneous dyads is most especially influenced by the complexity of the domain. In these microcells of collaborative dyads and groups, communication becomes particularly important, as it constructs the context and its dynamics. Exploring various consumer domains with varying complexity and a focus on communication is an insightful research path. The active integration of key individuals of the organization either in dyads or bigger groups marks another interesting path of research. Their unique information contains knowledge such as what kinds of ideas go well within the organization or what kinds of ideas have been rewarded in the past (Sternberg, 1998). Their objectives are rather motivated by monetary maximization than by maximization of value.

4.7 Conclusion

Potential users are valuable partners in explorative activities towards new market opportunities. This study showed how their merit is embodied in communication in seven categories and propositions, depicted their interrelation and identified beneficial antecedents. In active participation a potential user's virtue is constituted by what emerges from the situated context, with varying degrees of predictability. Their enrichment of the creative process and its content invokes a more subtle kind of thinking. Organizational empathy embedded in the skills of key individuals is indispensable in this context. The in-depth inquiry of potential users' contributions is crucial for practitioners and researchers in innovation management. The findings advocate a contextual stance of creative problem-solving and a focus on the research of communicative action in these contexts. Guidelines were depicted to support the creation of methods seeking to formalize organizational empathy and to mitigate the multitude of struggles which organizations face when exploring new markets. Tolerating unpredictability and improvisation are key factors therein.

References

Alba, J. W., & Hutchinson, W. J. 1987. Dimensions of Consumer Expertise. *Journal of Consumer Research,* 13(4): 411–454.

Amabile, T. M. 1983. The Social Psychology of Creativity: A Componential Conceptualization. *Journal of Personality and Social Psychology,* 45: 357–376.

Anthony, S. D. 2009. Major League Innovation. *Harvard Business Review,* 87(10): 51-54.

Archer, M. 1995. *Realist Social Theory: The Morphogenetic Approach.* New York: Cambridge University Press.

Arnold, T. J., Fang, E., & Palmatier, R. W. 2011. The Effects Of Customer Acquisition and Retention Orientations on a Firm's Radical and Incremental Innovation Performance. *Journal of the Academy of Marketing Science,* 39(2): 234-251.

Baker-Sennett, J., Matusov, E., & Rogoff, B. 1992. Sociocultural Processes of Creative Planning in Children's Playcrafting. In P. Light & G. Butterworth (Eds.), *Context and Cognition: Ways of Learning and Knowing:* 93-114. New York: Harvester-Wheatsheaf.

Bearison, D. J., & Dorval, B. 2002. *Collaborative Cognition: Children Negotiating Ways of Knowing.* Westport, CT: Ablex Publishing.

Bharadwaj, N., Nevin, J. R., & Wallman, J. P. 2012. Explicating Hearing the Voice of the Customer as a Manifestation of Customer Focus and Assessing its Consequences. *Journal of Product Innovation Management,* forthcoming.

Birkinshaw, J., Bessant, J., & Delbridge, R. 2007. Finding, Forming, and Performing: Creating Networks for Discontinuous Innovation. *California Management Review,* 49(3): 67-84.

Bogers, M., Afuah, A., & Bastian, B. 2010. Users as Innovators: A Review, Critique, and Future Research Directions. *Journal of Management,* 36: 857–875.

Bond, E., & Houston, M. 2003. Barriers to Matching New Technologies and Market Opportunities in Established Firms. *Journal of Product Innovation Management,* 20(2): 120-135.

Bonner, J. M., & Walker Jr., O. C. 2004. Selecting Influential Business-to-Business Customers in New Product Development: Relational Embeddedness and Knowledge Heterogeneity Considerations. *Journal of Product Innovation Management,* 21: 155-169.

Bruner, J. 1990. *Acts of Meaning.* Cambridge, MA: Harvard University Press.

Campbell, D. T. 1960. Blind variation and selective retention in creative thought as in other knowledge processes. *Psychological Review,* 67: 380–400.

Chandy, R. K., & Tellis, G. J. 1998. Organizing for Radical Product Innovation: The Overlooked Role of Willingness to Cannibalize. *Journal of Marketing Research,* 35(4): 474–487.

Christensen, C. M. 1997. *The Innovator's Dilemma.* Boston: Harvard Business School Press.

Christensen, C. M. 2006. The Ongoing Process of Building a Theory of Disruption. *Journal of Product Innovation Management*, 23: 39-55.

Christensen C. M., & Bower, J. L. 1996. Customer Power, Strategic Investment, and the Failure of Leading Firms. *Strategic Management Journal*, 17(3): 197–218.

Cole, M. 1996. *Cultural Psychology: A Once and Future Discipline.* Cambridge, MA: Belknap Press.

Crossan, M. M. 1998. Improvisation in Action. *Organization Science*, 9(5): 593-599.

Cunha, M. P., Cunha, V. J., & Kamoche, K. 1999. Organizational Improvisation: What, When, How and Why. *International Journal of Management Reviews*, 1: 299–341.

Danneels, E. 2002. The Dynamics of Product Innovation and Firm Competences. Strategic Management Journal, 23: 1095-1121.

Danneels, E. 2003. Tight-Loose Coupling with Customers: The Enactment of Customer Orientation. *Strategic Management Journal*, 24: 559-576.

Day, G. S. 1999. Misconceptions about Market Orientation. *Journal of Market Focused Management*, 4: 5-16.

Denzin, N. K. 1989. *The Research Act: A Theoretical Introduction to Sociological Methods.* Third edition, New York.

Dunbar, K. 1997. How Scientists Think: On-Line Creativity and Conceptual Change in Science. In T. B. Ward, S. M. Smith & J. Vaid (Eds.), *Conceptual Structures and Processes: Emergence, Discovery, and Change:* 461-493: Washington D.C: American Psychological Association Press.

Edmondson, A. 1999. Psychological Safety and Learning Behavior in Work Teams. *Administrative Science Quarterly*, 44(2): 350-383.

Eisenhardt, K. M., & Tabrizi, B. N. 1995. Accelerating Adaptive Processes: Product Innovation in the Global Computer Industry. *Administrative Science Quarterly*, 40: 84-110.

Faems, D., van Looy, B., & Debackere, K. 2005. Interorganizational Collaboration and Innovation: Toward a Portfolio Approach. *Journal of Product Innovation Management*, 22: 238–250.

Festinger, L. 1957. *A Theory of Cognitive Dissonance.* Stanford, CA: Stanford University Press.

Florén, H., & Frishammar, J. 2012. From Preliminary Ideas to Corroborated Product Definitions: Managing the Front End of New Product Development. *California Management Review*, 54(4): 20-43.

Frensch, P. A., & Sternberg, R. J. 1989. Expertise and Intelligent Thinking: When is it worse to know better? In R. J. Sternberg (Ed.), *Advances in the psychology of human intelligence,* vol. 5: 157-158. Hillsdale, NJ Erlbaum.

Fuchs, C., & Schreier, M. 2011. Customer Empowerment in New Product Development. *Journal of Product Innovation Management*, 28(1): 17-32.

Gephart, R. P. 2006. From the Editors: Qualitative Research and the Academy of Management. *Academy of Management*, 47(4): 454-462.

Given, L. M. (Ed.) 2008. *The Sage Encyclopedia of Qualitative Research Methods,* vol. 2: 697-698. Sage: Thousand Oaks, CA.

Glaser, B. G. 1978. *Theoretical Sensitivity.* The Sociology Press, San Francisco.

Glaser, B. G., & Strauss, A. L. 1967. *Discovery of Grounded Theory: Strategies for Qualitative Research.* Chicago: Aldine.

Glaveanu, V. P. 2010. Creativity as Cultural Participation. *Journal for the Theory of Social Behavior,* 41: 48-67.

Glaveanu, V. P. 2011. How are We Creative Together? Comparing Sociocognitive and Sociocultural Answers. *Theory and Psychology,* 21(4): 473-492.

Goldenberg, J., Mazursky, D., & Solomon, S. 1999. Toward Identifying the Inventive Tem plates of New Products: A Channelled IdeationApproach. *Journal of Marketing Research,* 36, 200–210.

Govindarajan, V., & Kopalle, P. K. 2006. Disruptiveness of Innovations: Measurement and an Assessment of Reliability and Validity. *Strategic Management Journal,* 27: 189-199.

Govindarajan, V., Kopalle, P. K., & Danneels, E. 2011. The Effects of Mainstream and Emerging Customer Orientations on Radical and Disruptive Innovations. *Journal of Product Innovation Management,* 28(S1): 121-132.

Greer, C. R., & Lei, D. 2012. Collaborative Innovation with Customers: A Review of The Literature and Suggestions for Future Research. *International Journal of Management Reviews,* 14(1): 63–84.

Guilford, J. P. 1956. The Structure of Intellect. *Psychological Bulletin,* 53(4): 267-293.

Guilford, J. P. 1968. *Creativity, Intelligence, and their Educational Implications.* San Diego, CA.

Hang, C., Chen, J., & Subramian A. M. 2010. Developing Disruptive Products for Emerging Economies: Lessons from Asian cases. *Research Technology Management:* 21-26.

Hauser, J. R., Tellis, G. J., & Griffin, A. 2006. Research on Innovation: A Review and Agenda for Marketing Science. *Marketing Science,* 25(6): 687-717.

Henard, D. H., & Szymanski, D. M. 2001. Why Some New Products are More Successful than Others. *Journal of Marketing Research,* 38(3): 362-375.

Henderson, R. 2006. The Innovator's Dilemma as a Problem of Organizational Competence. *Journal of Product Innovation Management,* 23: 5-11.

Hewing, M. 2013. Merits of Collaboration with Potential and Current Users In Creative Problem-Solving. *International Journal of Innovation Management,* 17(3):1-27.

Hinsz, V.B., Tindale, R.S., & Vollrath, D.A. 1997. The Emerging Conceptualization of Groups as Information Processors. *Psychological Bulletin,* 121(1): 43-64.

Hutchins, E. 1995. *Cognition in the Wild.* Cambridge, MA: MIT Press.

Hymes, D. 1974. *Foundations in Sociolinguistics.* London: Tavistock Publications Limited.

Johnson, R. B., Onwuegbuzie, A. J., & Turner, L. A. 2007. Toward a Definition of Mixed Methods Research. *Journal of Mixed Methods Research,* 1(2): 112-133.

Jovchelovitch, S. 2007. *Knowledge in Context: Representations, Community and Culture.* London, UK: Routledge.

Koestler, A. 1966. *Der göttliche Funke: Der schöpferische Akt in Kunst und Wissenschaft.* Wien und München: Scherz Verlag.

Kratzer, J., & Lettl, C. 2008. A Social Network Perspective of Lead Users and Creativity: An Empirical Study among Children. *Creativity and Innovation Management,* 17(1): 26-36.

Kristensson, P., Matthing, J., & Johansson, N. 2008. Key Strategies for the Successful Involvement of Customers in the Co-Creation of New Technology-Based Services. *International Journal of Service Industry Management,* 19(3-4): 474-491.

Kyriakopoulos, K. 2011. Improvisation in Product Innovation: The Contingent Role of Market Information Sources and Memory Types. *Organization Studies,* 32(8): 1051–1078.

Larey, T., & Paulus, P. 1999. Group Preference and Convergent Tendencies in small Groups: A Content Analysis of Group Brainstorming Performance. *Creativity Research Journal,* 12(3): 175–184.

Lau, A. K. W., Tang, E., & Yam, R. C. M. 2010. Effects of Supplier and Customer Integration on Product Innovation and Performance: Empirical Evidence in Hong Kong Manufacturers. *Journal of Product Innovation Management,* 27(5): 761-777.

Leonard-Barton, D. 1992. Core Competencies and Core Rigidities: a Paradox in Managing New Product Development. *Strategic Management Journal,* 13: 111–125.

Leonard, D., & Sensiper, S. 1998. The Role of Tacit Knowledge in Group Innovation. *California Management Review,* 40: 112–132.

Litchfield, R. C. 2008. Brainstorming Reconsidered: A Goal-Based View. *Academy of Management Review,* 33(3): 649–668.

Mascitelli, R. 2000. From Experience: Harnessing Tacit Knowledge to Achieve Breakthrough Innovation. *Journal of Product and Innovation Management,* 17: 179-193.

Meyer, A. D. 1982. Adapting to Environmental Jolts. *Administrative Science Quarterly,* 27(4): 515-537.

Miles, M. B., & Huberman, A. M. 1994. *Qualitative Data Analysis: An Expanded Sourcebook.* Thousand Oaks: Sage Publications.

Mills, P. K., Chaes, R., & Margulies, N. 1983. Motivating the Client/Employee System as a Service Production Strategy. *Academy of Management Review,* 8: 301–310.

Miner, A. S., Bassoff, P., & Moorman, C. 2001. Organizational Improvisation and Learning: A Field Study. *Administrative Science Quarterly,* 46(2): 304-337

Moorman, C., & Miner, A. S. (1998). Organizational Improvisation and Organizational Memory. *Academy of Management Review,* 23: 698–723.

Mumford, M. D. 2003. Where have We been, here are We Going? Taking Stock of Creativity Research. *Creativity Research.Journal,* 15: 107–120.

Osborn, A. F. 1963. *Applied Imagination: Principles and Procedures of Creative Problem-solving.* New York, NY: Charles Scribner's Sons.

Paulus, P. B., & Nijstad, B. A. 2003. Group Creativity. In P. B. Paulus & B. A. Nijstad (Eds.), *Group Creativity:* 3-11. Oxford University Press.

Paulus, P. B., & Yang, H. C. 2000. Idea Generation in Groups: A Basis for Creativity in Organizations. *Organizational Behavior and Human Decision Processes,* 82: 76-87.

Prahalad, C. K., & Ramaswamy, V. 2004. Co-creating Unique Value with Customers. *Strategy and Leadership,* 32(3): 4–9.

Reid, S. E., & Brentani, U. 2004. The Fuzzy Front End of New Product Development for Discontinuous Innovations: A Theoretical Model. *Journal of Product Innovation Management,* 21: 170-184.

Rogoff, B. 1995. Observing Sociocultural Activity and Three Planes: Participatory Appropriation, Guided Participation, and Apprenticeship. In J.V. Wertsch, P. del Rio & A. Alvarez (Eds.), *Sociocultural Studies of Mind:* 139-164. New York: Cambridge University Press.

Rosenberg, M. 1956. Cognitive Structure and Attitudinal Affect. *Journal of Abnormal and Social Psychology,* 53: 367-72.

Rubenson, D. L., & Runco, M. A. 1995. The Psychoeconomic View of Creative Work in Groups and Organizations. *Creativity and Innovation Management,* 4: 232-241.

Runco, M. A., & Sakamoto, S. O. 1999. Experimental Studies of Creativity. In R. J. Sternberg (Ed.), *Handbook of Creativity:* 62-92. New York: Cambridge University Press.

Saville-Troike, M. 2003. *The Ethnography of Communication.* Oxford: Blackwell Publishing.

Sawyer, R. K. 2005. *Social Emergence.* Cambridge University Press, Cambridge.

Sawyer, R. K. 2006. *Explaining Creativity.* Oxford University Press, New York.

Shaw, B. S. 1985. The Role of the Interaction between the User and the Manufacturer in Medical Equipment Innovation. *R&D Management,* 15: 283-292.

Shweder, R. 1990. Cultural Psychology – What is it? In J. Stigler, R. Shweder, & G. Herdt (Eds.), *Cultural Psychology: Essays on Comparative Human Development,* 1–43: Cambridge, UK: Cambridge University Press.

Simonton, D. K. 1977. Creative Productivity, Age and Stress: A Biographical Time-Series Analysis of 10 Classical Composers. *Journal of Personality and Social Psychology,* 35: 805–816.

Slater, F. S., & Mohr, J. J. 2006. Successful Development and Commercialization of Technological Innovation: Insights Based on Strategy Type. *Journal of Product Innovation Management,* 23: 26-33.

Smith, S., Gerkens, D., Shah, J., & Vargas-Hernandez, N. 2006. Empirical Studies of Creative Cognition in Idea Generation. In L. Thompson & H.-S. Choi (Eds.), *Creativity and Innovation in Organizational Teams,* 3–20: Mahwah, NJ: Lawrence Erlbaum Associates.

Sonnenburg, S. 2004. Creativity in Communication: A Theoretical Framework for Collaborative Product Creation. *Creativity and Innovation Management,* 13(4): 254–262.

Sternberg, R. J. 1998. Cognitive Mechanisms in Human Creativity: Is Variation Blind or Sighted? *Journal of Creative Behavior,* 32: 159-176.

Steyaert, C., & Bouwen, R. 2004. Group Methods of Organizational Analysis. In C. Cassell & G. Symon (Eds.), *Essential Guide to Qualitative Methods in Organizational Research:* 140-153. London: Sage.

Strauss, A., & Corbin, J. 1996. *Grounded Theory.* Weinheim: Beltz.

Stringer, R. 2000. How to Manage Radical Innovation. *California Management Review,* 42:70-88.

Suddaby, R. 2006. What Grounded Theory is Not. *Academy of Management,* 49(4): 633-642.

Teddlie, C., & Tashakkori, A. 2011. Mixed Methods Research: Contemporary Issues in an Emerging Field. In N. K. Denzin, & Y. S. Lincoln (Eds.), *The Sage Handbook of Qualitative Research (4th ed.):* 285-299: Sage.

Ulwick, A. W. 2002. Turn Customer Input into Innovation. *Harvard Business Review:* 91-97.

van der Panne, G., van Beers, C., & Kleinknecht, A. 2003. Success and Failure in Innovation: A Literature Review. *International Journal of Innovation Management,* 7: 309–338.

Vargo, S. L., & Lusch, R. F. 2004. Evolving to a New Dominant Logic for Marketing. *Journal of Marketing,* 68(1): 1-17.

Vera, D., & Crossan, M. 2005. Improvisation and Innovative Performance in Teams. *Organization Science,* 16(3): 203–224

von Hippel, E. 1986. Lead Users: A Source of Novel Product Concepts. *Management Science,* 32: 791–805.

von Hippel, E. 1994. Sticky Information and the Locus of Problem-solving: Implications for Innovation. *Management Science,* 40: 429-439.

Voss, J. F., & Post, T. A. 1988. On the Solving of Ill-Structured Problems. In M. Chi, R. Glaser & M. Farr (Eds.), *The Nature of Expertise:* 261-285. Hillsdale.

Ward, T. B. 1995. What's old about new Ideas? In S. M. Smith, T. B. Ward & R. A. Finke (Eds.), *The Creative Cognition Approach:* 157-178. Cambridge, MA: MIT Press.

Ward, T. B. 2007. Creative Cognition as a Window on Creativity. *Methods,* 42: 28-37.

Wiley, J. 1998. Expertise as mental set: the effects of domain knowledge in creative problem-solving. *Memory & Cognition,* Vol. 26 No. 4: 716-30.

Weisberg, R. W. 1999. Creativity and Knowledge: A Challenge to Theories. In R.J. Sternberg (Ed.), *Handbook of Creativity:* 226-250. New York: Cambridge University Press.

Zaichkowsky, J. L. 1985. Measuring the Involvement Construct. *The Journal of Consumer Research,* 12: 341-352.

5 Appendix: A Theoretical and Empirical Comparison of Innovation Diffusion Models[11]

Martin Hewing

Abstract This paper presents a theoretical and empirical comparison of quantitative innovation diffusion models with data from the software industry. Looking at some of the most widely used aggregate models, such as the Generalised Bass and the Kalish model, the inclusion of decision variables is compared to scientific groundings from innovation marketing, focusing on the analysis of their mathematical equations. Eligible attributes such as the reduction and the carry-through effect are addressed and performance simulations of the mapping function are run, using commonly observed pricing strategies. Adoption and marketing data of two innovative software products are applied to calibrate the models and to evaluate their forecasting precision by comparing the results with true data within a period of 10 months after initial release. It becomes evident that the predominant S-shaped models are not always suitable. A market analysis shows the complexity of market structures and the involved requirements for these models. Critical branch-related aspects, such as the reputation of the company, online distribution channels, and piracy are discussed and disclose future research spots in the estimation of diffusion shapes of innovative products.

5.1 Introduction

This paper examines and applies two of the most acknowledged diffusion models - the Generalised Bass (1994) and the Kalish (1985) model. These models structure and analyse the diffusion process of innovative products and try to forecast the diffusion trajectory over the product's lifetime. The research focus involves the integration of marketing mix variables into these models, in particular rollout price and marketing expenses. The contribution to the literature is an examination of the mathematical functions of these models, their impact on the diffusion trajectory and a comparison to theories from marketing, focusing on interdependencies of diffusion parameters and decision variables. Therefore, simulations of common pricing strategies will be run to analyse whether the models accurately reflect their dynamic changes. Afterwards, adoption and marketing data from two innovative software products will be applied to

---

[11] This chapter was previously published in the Journal of the Knowledge Economy, vol. 3(2), © Springer Science+Business Media, LLC 2011.

calibrate the models and evaluate their forecasting precision by comparing the results with true data within a period of 10 months after initial release. Actual application of diffusion models to true data is extremely scarce and constitutes the major contribution of this paper. It sheds light on actual application of the mathematical functions and benefits and obstacles when using its results for normative inferences. A market analysis will show the complexity of market structures of the software branch and the involved requirements for these models. Critical branch-related aspects, such as the reputation of the company, online distribution channels, and piracy are discussed and disclose future areas of research in the estimation of diffusion shapes of innovative products.

5.2 Diffusion Models with Decision Variables

Models with constant parameters are criticized because they neglect dynamics decision variables exert on diffusion processes (Dockner & Jorgensen, 1988). Every empirical adoption curve is associated to a sequence of prices and other factors, such as promotion activities (Mahajan et al., 1995). Therefore, it is reasonable to integrate economic variables into the modelling. While the Bass model (1969) concentrates on modelling with variables from the demand side, recent publications incorporate variables from the supply side (Bottomley & Fildes, 1998). Consequently, these models should focus on effects of different marketing mix strategies relating to diffusion curve and timing of peak sale (Dockner et al., 1988). Furthermore, these models can be used to support the detection of optimal strategies by numerical optimization (Kalish & Sen, 1986). This would address the normative appliance of these models. Since the impact of these variables is highly complex, there is no common guiding theory (Dockner et al., 1988). This has led to a multitude of models over the last 30 years.

5.2.1 The Role of Price

Two different views have emerged within the research of innovation diffusion, both regarding the relation between price changes and market potential. One approach argues that a price change influences the market potential (Chow, 1967; Mahajan & Peterson, 1978; Kalish, 1985). A price reduction would place the innovation within the budget limits of a growing number of individuals and thus increases the absolute market potential. The other approach states that the market potential remains constant while the probability of adoption changes (Bass, 1980). A price reduction would cause an advanced adoption of potential users. Both views can be substantiated by theories of behavioural science. Kamakura and Balasubramanian (1988) did research on both hypotheses based on the market segmentation that Mahajan and Muller (1979) suggested. The concept of reservation price supports both hypotheses. However, Kamakura and Balasubramanian (1988) stated that especially with high price innovations, the diffusion process is influenced by the adoption probability. Nevertheless, the theory of a static market potential is not very convincing (Mahajan et al., 1990). The ex-ante quantification of the market size is a major point of critique within the modelling of diffusion (Raman & Chatterjee, 1995). To understand the price effect in more detail, it is crucial to consider the impact of mathematical incorporations of price effectiveness into the equation and find out whether this is in line with the argumentation of price impact theories.

5.2.2 The Role of Advertisement

Advertisement communicates information. An on-going research discussion deals with the influence of information on individuals' decision heuristics (Kalish et al., 1986). Yet, the literature agrees on the notion that advertisement raises awareness and communicates the essential characteristics of an innovation. In the context of diffusion theory, an essential requirement for adoption is the social system's awareness of the existence of an innovation. Due to its range, advertisement is an adequate means to increase awareness of an innovation efficiently and fast. Models incorporating advertisement often use a stimulus–response function that exerts influence on the adoption.

5.2.3 The Role of Shadow Diffusion

Since the focus of the empirical analysis will be on software products, shadow diffusion needs to be discussed. Shadow diffusion is the illegal diffusion of software products in the marketplace caused by software piracy (Givon et al., 1995). It is highly complex to incorporate piracy in the modelling of diffusion, since it does not only lower the market potential in general. Just like legal diffusion, the diffusion of piracy copies is a dynamic process. An illegal version becomes available after (or sometimes even before) rollout and expatiates upon society. On the one hand, it is used by individuals who otherwise would have bought a copy of the product. On the other hand, it is used by individuals who only use the product because they do not have to pay for it. The system of supply and demand associated with a market price is blurring. However, there is also an upside associated with piracy. Illegal adopters play an important role in the adoption process of legal adopters. They serve as transmitters of mouth-to-mouth communication. Furthermore, illegal adopters can turn into legal adopters. Givon et al. (1995) proposed a model that incorporates shadow diffusion and its reciprocity with the legal diffusion.

5.2.4 Desired Characteristics of Models with Decision Variables

Bass et al. (2000) discussed diffusion models with decision variables. They proposed a number of desired characteristics for modelling with decision variables. Some of them will be used for the present discussion. These characteristics are *reduction, carry through effect, empirical support, analytical solution, applications in decision processes,* and *estimation without data.*

5.3 The Generalised Bass Model

Bass et al. (1994) expand the basic Bass model (1969, BM) and add the so-called mapping function. While the BM imputes an automatism to the diffusion process, the Generalised Bass model (GBM) uses the mapping function to incorporate decision variables, while conserving the essential characteristics of its origin (referring to the desired characteristic of reduction). Conditioned by the choice of decision variables within the mapping function, the date and height of the peak sale differs (Mahajan et al., 1995). Bass et al. (1994) consider price and advertisement as decision variables. The GBM is then presented by (1):

$$\frac{f(t)}{[1 - F[T]]} = [p + qF(T)] \cdot [1 + [\frac{Pr'(t)}{Pr(t-1)}] \cdot \beta_1 + [\frac{Adv'(t)}{Adv(t-1)}] \cdot \beta_2] \qquad (1)$$

$f(t)$	Density function
$F(t)$	Distribution function
p	Innovation coefficient
q	Imitation coefficient
$Pr(t)$	Price at time t
$Pr'(t)$	Price alteration at time t, $Pr(t) - Pr(t-1)$
$Adv(t)$	Advertisement spending at time t
$Adv'(t)$	Advertisement spending alteration at time t
β_1	Effectivity of prices
β_2	Effectivity of advertisement spending

Coefficient β_1 and β_2 represent the effectiveness of price and advertisement. Both can accelerate or delay the diffusion process. The assumptions here are a negative algebraic sign for price effectiveness β_1 and a positive sign for effectiveness of advertisement β_2. Price and advertisement of one period are related to their parameter value of the preceding period, so the notion of marginal decreases in gains is respected. Alterations (e.g. increase of advertisement spending) affect the probability of adoption less if the base level increases.

5.3.1 GBM Simulation of Common Pricing Strategies

Regarding the changes in price, the adoption theory supplies the hypothesis that the preceding stimulus - the price of the preceding period - is the benchmark to which customers compare the current price (Srinivasan & Mason, 1986). The mapping function of the GBM supports this theory (s.a.). However, it is still interesting to see how adoption adapts to price changes in time elapse. Does it reflect dynamic changes of pricing strategies? To shed light on this question, three different commonly acknowledged pricing strategies will be used. In the following, a constant, a volatile, and an exponentially decreasing pricing strategy are modelled. Figure 5.A presents the pricing strategies and the respective diffusion trajectories on a time line. The trajectories portrayed are based on a simulation of the GBM with given diffusion parameters. The advertisement parameter was kept constant for this simulation. **Constant price.** Holding the price constant over all periods renders the price parameter zero. The price level only exerts influence in the first period of adoption, since by definition the price in $Pr(0) = 0$. Consequently, there is only one alteration of the price level, between $t = 0$ and $t = 1$, which equals the price tag. Figure 5.A shows that the diffusion reaches the peak sale rather late when the price is kept constant. The price tag of the constant price strategy is located in the middle of the other strategies' extremes. The constant price strategy has the highest average price in this simulation. Therefore, a delayed diffusion is reflected properly.

Volatile price. The volatile price strategy exerts an influence similar to the constant price strategy. The simulation supposes that the price changes in one period only, so the alteration is influential in that particular period only. During the other periods, the mapping function renders the price parameter to zero. In reality, volatile pricing strategies are often associated with durable consumer goods (Simon, 1992).

Figure 5.A: Diffusion Simulation along three Pricing Strategies

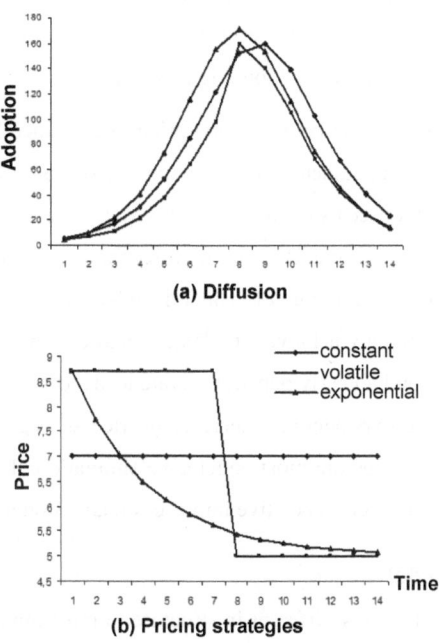

(a) Diffusion

(b) Pricing strategies

Goods with high technology dynamics often have high obsolescence rates and therefore they have high price tags at first. Regarding real innovations, monopoly allows for such strategy. Over time, the product loses its novelty and prices are reduced drastically (Simon, 1992). In the function of the GBM model, the price alteration only impacts during the period of change, ignoring the carry-through effect. A steady low price would suggest a long time effect on the adoption probability. Price reductions and increases do not have symmetrical influence. The model, however, assumes a symmetrical influence. A reduction and an increase would have the same influence on adoption, each with a reciprocal algebraic sign. The prospect theory supposes an asymmetry of the impact. A loss would affect the market participant significantly more than a gain of the same amount (Kahneman & Tversky, 1979). If you equalise a price increase with a loss and a price reduction with a gain, this would mean, that the price elasticity is higher with regards to a price increase than to a price reduction (Simon, 1992). Figure 5.A shows the effect of fitful mitigation on price explicitly in period seven. A longer-lasting effect on subsequent periods, however, does not arise. The integration of price elasticity into the mapping function might be fruitful.

Exponentially decreasing price. In reality, prices are frequently reduced in a constant percental manner. Many manufacturers of consumer goods adjust their pricing strategies to dynamics of production costs (Simon, 1992). Experience curve effects render steady cost reductions possible - especially in high-technology branches. Then, the price management bases on the relation of demand and cost dynamics over time. In the simulation, this price strategy achieves the fastest market penetration - due to relatively early price reduction - and an early penetration rate leads to wide dispersion of mouth-to-mouth information. The exponentially decreasing price strategy has the lowest average price over time and achieves the most conclusive simulation. The reason for this is the continuity of reduction that gives positive impulses on adoption in each period.

5.4 The Kalish Model

The Kalish model (1985) is different in structure to most other diffusion models. It is not based on the BM's mathematical structure and therefore interesting to analyse. The diffusion process is not binary. While the BM acts on the assumption that potential customers either adopt or reject the innovation (Mahajan et al., 1990) (so no stages of adoption process are considered), Kalish distinguishes between the diffusion of information and the actual purchase of an innovation. Gatignon and Robertson (1991) name three constructs that drive diffusion and can be influenced by marketing-mix variables: the awareness, the willingness to pay the market price, and the fact that the product is available. The willingness to pay is dependent on individual utilities and uncertainty. Kalish's model reflects the awareness and the willingness to pay.

The mathematical conversion of the process is done with two separate equations:

Awareness: $I(t) = [1 - I] \cdot [f(A(t)) + bI + b'\left(\dfrac{X}{N_0}\right)]$ \hfill (2)

I	Fraction of the product-aware part of population at time t
X	Adopter at time t
N_0	Market potential, equates to amount of adopters at price of zero
$A(t)$	Advertisement spending at time t
$f(.)$	Effectivity of advertisement spending
b	Influence of aware individuals
b'	Influence of the adopters

In this model, advertisement shapes awareness and defines thereby the market potential as reflected in Equation 2. This means that the model assumes a dynamic market potential. The actual adoption is reflected in (3):

Adoption: $[N_0\left(\dfrac{p}{u(\frac{X}{N})}\right) \cdot I - X] \cdot k$ \hfill (3)

p	Price at time t
u	Discount factor

Note that the outcome of Equation 2 is integrated into Equation 3. The adopting probability is dependent on individual utilities that are brought into the model via the concept of reservation price. Furthermore, the reservation price is discounted by the factor u which represents the individual uncertainty associated with the utility. The process of adoption in the diffusion literature is characterised by a *hierarchy of effects* model (Gatignon et al., 1991). An individual runs through a sequential process of change: from the change of knowledge to the change in behaviour. Rogers (1995) supposes five steps of this process, with awareness being the first one. The influence of mass media is emphasised at this point. The diffusion literature divides the use of communication channels along the adoption process. Thus, mass media is critically important in the early phases of creating awareness while interpersonal communication determines later stages

(Gatignon et al., 1991). Kalish picks up the sequential nature in his mathematical conversion, regarding adoption as a process.

5.4.1 Kalish Simulation of Common Pricing Strategies

Many diffusion models regard the market potential as a constant that is defined at the beginning of the diffusion process (Mahajan et al., 1990). This assumption is escapist, especially in context of innovation where predictions of acceptance and potential are always uncertain. Kalish, however, incorporates a dynamic market potential into his model and follows a far more realistic modelling. With regards to the segments defined by Mahajan and Muller (1979), Kalish assumes that a price reduction has an expanding effect on the market potential. After testing the impact of different pricing strategies on the probability of adoption, it is interesting to test the impact of different pricing strategies on the market potential that influences the adoption rate indirectly.

Figure 5.B: Adaptor Segments according to Mahajan et al. (1979)

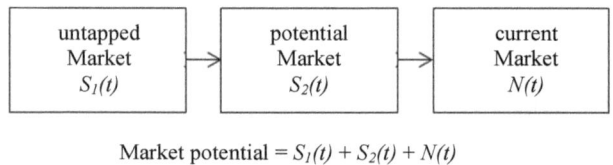

Market potential = $S_1(t) + S_2(t) + N(t)$

The untapped market, as shown in Figure 5.B, consists of individuals whose adjusted reservation price does not exceed the market price of the innovation. Figure 5.C shows how different pricing strategies influence the market potential at different periods and how this influences the diffusion process. The pricing strategies equal the simulation for the GBM and advertisement is again kept constant. Figure 5.C shows explicitly how pricing strategies foster different market dynamics. The constant and the exponential pricing strategies trigger a similar diffusion process in the first six periods, while constant prices lead to earlier market saturation. The process reaches the peak sale earlier and at a comparably lower level. Similar to the GBM, the brisk price reduction of the volatile price strategy can be noticed.

Figure 5.C: Diffusion Simulation under Market Dynamics

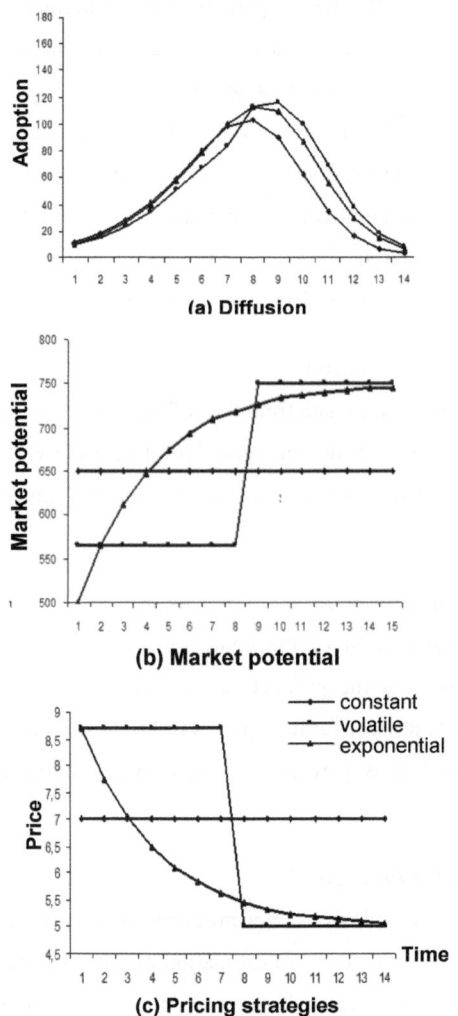

(a) Diffusion

(b) Market potential

(c) Pricing strategies

Between period 7 and 8, the adoption is taking off again before it seems to have reached its peak sale by lower marginal gain in the preceding period. The simulation of the model displays the effects of varying pricing strategies in a plausible manner. In reality, however, the price is not the only influencing factor of market potential. Kalish himself criticised the fact that the market potential is only defined by the price (Kalish, 1985). The market potential of one provider is always conditioned to the competition. In the context of real innovations, an initial monopoly position is often observed at the time of rollout (Simon, 1992). During the product life cycle these conditions change dramatically. A company is opposed to competitors and consumers do not only have the choice between adoption or non-adoption, but also between different suppliers. The diffusion literature offers different approaches to include the competitor situation into the modelling (Teng & Thompson, 1983; Rao & Bass, 1985).

5.5   Market Analysis and Method

Now both models are applied to data from the software industry. Furthermore, the data is also applied to the BM to see if the incorporation of decision variables, as done in the model of Kalish and the GBM, achieves better results than in a simpler model (BM).

5.5.1   Data Basis

The data (sales of 10 months, price and marketing expenses) is from two products of a well-known software provider. Since the company needs to be kept anonymous, the novelty and innovativeness of the products cannot be displayed here. However, it is assured that both products at least fulfil Rogers' (1995) definition: "An innovation is an idea, practice, or object that is perceived as new by an individual or other unit of adoption."

5.5.2   The Longevity of the Products

Both models focus on durable consumer products. Both considered products can be seen as durables, even though there is no estimation of their general life cycle since the company works constantly on updates that improve the products. Updates can be free of charge or liable to pay costs, depending on the extent of the update. Succeeding generations are mostly liable to pay costs and cannibalise precedent generations. For product 1, there have been two gratis updates in the considered period of time. There has not been a changeover in the generation though, so cannibalising effects do not need to be considered. For product 2, no updates were released in the considered time span.

5.5.3 Availability and Distribution

Since the diffusion literature assumes steady availability of the regarded objects, excess demands are not considered. In reality, however, gaps in supplies occurs quiet often. Jain et al. (1991) present how the diffusion process is influenced by these gaps. In the context of software products, a steady availability is quiet intuitive, since there is no physic product associated with it and copies can be produced easily. Product 1 is delivered with a piece of hardware that is produced by an outsourced entity. However, there has been no data on gaps available.

5.5.4 Shadow Diffusion

Obviously, the gravest problem the software industry is facing today is the act of piracy. Piracy promotes the illegal diffusion of software parallel to its legal diffusion. There is no accurate estimation from the company, but executives estimate an illegal usage of 50–80% of all users. One strategy to embank piracy is to peg the software to a hardware piece and ship them together, as they do in the case of product 1. However, the software still works without the hardware, only its usability is increased explicitly in case you use it in conjunction with the hardware. The effect of the software–hardware combination on its diffusion is therefore ambiguous, since the hardware raises the market price of the product. A consumer's reservation price needs to be accordingly higher so that the net profit is positive. Due to the ambiguity, piracy aspects are not regarded in the estimation and constitute the biggest shortcoming.

5.5.5 Advertisement, Reputation and Diffusion Curves

Advertisement spending for both products is done late after product introduction (3-month in delay). So two questions arise: what has driven the diffusion process before and why are peak sales located directly after market introduction? It is assumed that the reputation of the company needs to be referenced in this aspect. The company is held in high esteem due to the steady high quality of their products. It has led them to a steady user base in the private as well as in the business area. A good reputation diminishes the risk of adoption of a potential adopter (Selnes, 1993). In this case, the diffusion trajectory often follows a decreasing course as shown in Figure 5.D. Horsky and Simon (1983) call it a *promising innovation*.

Figure 5.D: Adoption Curves according to Horsky and Simon (1983)

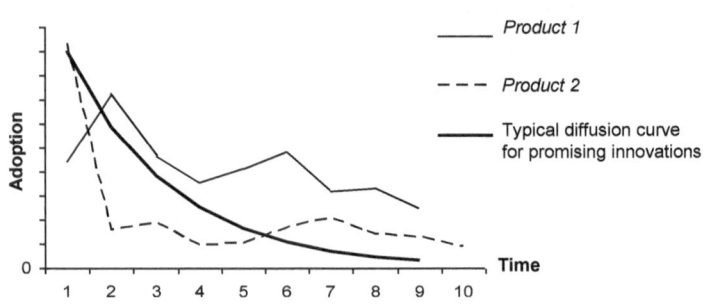

Furthermore, there are some effective advertising measures that are not included in the advertisement spending, such as a monthly product newsletter, reviews in international trade magazines, and the dissemination of information via their own internet page. Especially test reports trigger awareness and mouth-to-mouth communication. Within the estimation, these activities were considered within a dedicated parameter. The internet page offers gratis demo versions of the products. Demo versions are appropriate to reduce the consumer's risk associated with the adoption and increase the adoption speed and probability (Liftin, 2000).

5.6 Application of the Data to the Models

In the diffusion literature, the fit of the estimation to the real course of the diffusion curve, statistically represented by the parameter value of the coefficient of determination R^2, often it is the only evaluation criterion for the proposed model. Furthermore, by far not all new proposed models are empirically tested in the publications. The coefficient of determination thereby does not sufficiently express the performance of a model. Another criterion is the accuracy of its prediction. The fit is a quality criterion for the theoretical basis of the model and its descriptive application. The precision of prognosis tells you something about the applicability of the model in a normative way (Kalish et al., 1986). For the prognosis, the estimated diffusion parameters were substituted into the models. Afterwards, further adoption numbers for one subsequent period were computed, "running" the models in their calibration. These are compared to the real adoption data of the respective period. Furthermore, the parameter value of each estimation parameter are analysed with regards to plausibility. For estimation of the parameters of the models, the non-linear regression (NLS) was used. The NLS is seen as the method better suited with regards to biases and standard error than the OLS method (Srinivasan et al., 1986; Putsis & Srinivasan, 2000).

5.6.1 The Estimation

The equation of the GBM for its parameter estimations looks as follows:

$$s(t) = (p + \frac{q}{m} \cdot Y(t-1)) \cdot (m - Y(t-1)) * (1 + Pr'(t) \cdot \beta_1 + max\{0, Adv'(t)\} \cdot \beta_2) \quad (4)$$

$s(t)$	Adoption at time t
p	Innovation coefficient
q	Imitation coefficient
$Y(t)$	Cumulated adoption at time t
m	Absolute market potential
Pr'	Difference in price to preceding period in percent
β_1	Parameter of price alteration
$Adv'(t)$	Difference in advertisement spending to preceding period in percent
β_2	Parameter of advertisement spending alteration impact

In this model, the parameters p, q, m, $\beta 1$ and $\beta 2$ were estimated. The incorporation of the advertisement spending is done similar to the incorporation by Bass et al. (1994). Alterations in spending only exert an impact on adoption in case it is positive. The effect thereby has an asymmetric impact (Simon, 1982). Kalish's model needs an equation that unifies its two-stage process within one equation. Furthermore, assumptions on the distribution of the reservation price are needed, since information on individual reservation prices were not available. The estimation equation looks as follows:

$$s(t) = [z - k \cdot Pr(t) - Y(t-1)] \cdot [\beta_1 \cdot LN(Adv) + q \cdot Y(t-1) + \beta_3] \quad (5)$$

The following is an explanation of the derivation of the parameters z and k. A uniform distribution of reservation prices among the population was assumed. The reservation price therefore takes a certain value between the minimal (assumed to be zero) and the maximum possible value with the same probability. The probability that the reservation price has a value between the maximum price and the market price of the product specifies the fraction of the population that is willing to adopt. Their reservation price is equal or bigger than the market price of the product. Figure 5.E shows the density function of the distribution of the reservation price.

Figure 5.E: Density Function of the Individual Reservation Prices

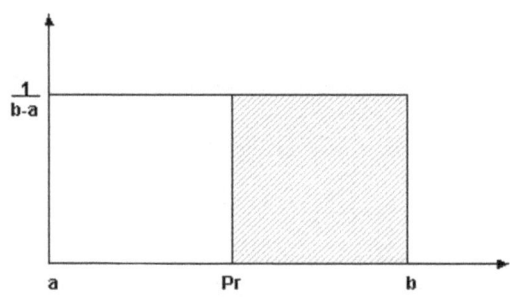

In case the market price is identical to a, the lowest reservation price, everybody would adopt the product. If you assign a market price above a, the fraction of adopters is defined by the area between the market price and the maximum reservation price of an individual b. If you multiply this fraction with the market potential at a market price of zero and subtract the cumulated adoption of the preceding periods in each period, you get the amount of adopters at time t. Mathematically it looks like this:

$$s(t) = \left[N_0 \left(\frac{b}{b-a} - \frac{1}{b-a} \cdot Pr(t) \right) - Y(t-1) \right] \cdot \left[\beta_1 \cdot LN(Adv) + q \cdot Y(t-1) + \beta_3 \right]$$

$$= \left[z - k \cdot Pr(t) - Y(t-1) \right] \cdot \left[\beta_1 \cdot LN(Adv) + q \cdot Y(t-1) + \beta_3 \right] \qquad (6)$$

$s(t)$	Adoption at time t
N_0	Absolute market potential at price 0
b	Highest reservation price of an individual
a	Lowest reservation price (assumed Value = 0)
Pr	Real market price at time t
$Y(t)$	Cumulated adoption at time t
β_1	Parameter of information diffusion by advertisement
Adv	Advertisement spending at time t
q	Parameter of information diffusion by mouth-to-mouth
β_3	Parameter of other communication activities
z	$N_0 * \dfrac{b}{b-a}$

$$k \qquad N_0 * \frac{1}{b-a}$$

The parameters z, k, $\beta 1$, q and $\beta 3$ needed estimation in this model. The parameter $\beta 3$ presents the communication activities mentioned above (Newsletter, Homepage, etc.). Regarding the GBM, this parameter did not lead to a better result of the estimation, so it was discarded.

5.6.2 Parameter Estimations

The estimators, the asymptotic standard error and the coefficient of determination for both products are shown in Table 5.A for product 1 and Table 5.B for product 2.

The Fit. The coefficient of determination R^2 of the estimations GBM and Kalish is apparently high. The models exhibit therefore a good fit to the real adoption data. Figure 5.F shows the real adoption together with the fit of the models over time. The BM has shortcomings, especially when trying to fit to the data of product 1. In Figure 5.F, it becomes clear that the adoption of product 1 varies heavily between periods.

The BM, with its supple curve, cannot reproduce these heavy oscillations consumption, and therefore has a high variance. The Kalish and the GBM models are more flexible due to the incorporation of marketing mix variables. Their curve follows the oscillations. The BM shows more of a tendency towards where the diffusion is heading. The GBM reaches a better fit regarding both products than the model of Kalish. For both products, advertisement spending was done only within a certain period of time after rollout. However, this information especially leads to a higher fit of the GBM. The model expansion is thereby justified and helpful. The adoption of product 2 does not vary as strong as the one of product 1. The BM therefore reaches a better fit here.

Table 5.A: Parameter Estimations - Product 1

	GBM	ASE	BM	ASE	Kalish	ASE
R^2	0,91		0,53		0,85	
p	0,04*	0,01*	0,10*	0,03*		
q	0,30*	0,05*	0,08	0,14	-0,00	0,00
m	4407*	276,7*	5223*	1894,56*	7184	
β_1	1,62*	0,47*				
β_2	0,06	0,06			0,01	0,07
β_3					0,11	0,43
N_0					803.767.472	1,08E+16

*significant estimator; GBM = Generalised Bass Model, ASE = asymptotic standard error, BM = Bass Model

Table 5.B: Parameter Estimations - Product 2

	GBM	ASE	BM	ASE	Kalish	ASE
R^2	**0,98**		**0,85**		**0,97**	
p	0,04	0,08	0,38	5641,7		
q	0,11	0,37	-0,38	5641,81	-0,00	0,00
m	3096	2336	2260	3,3E+07	2439	
β_1	7,41	24,56				
β_2	0,16	0,41			0,08	0,12
β_3					0,43*	0,17*
N_0					1.271.127	2,93E+13

* significant estimator

Figure 5.F: Real Adoption and Model Fit

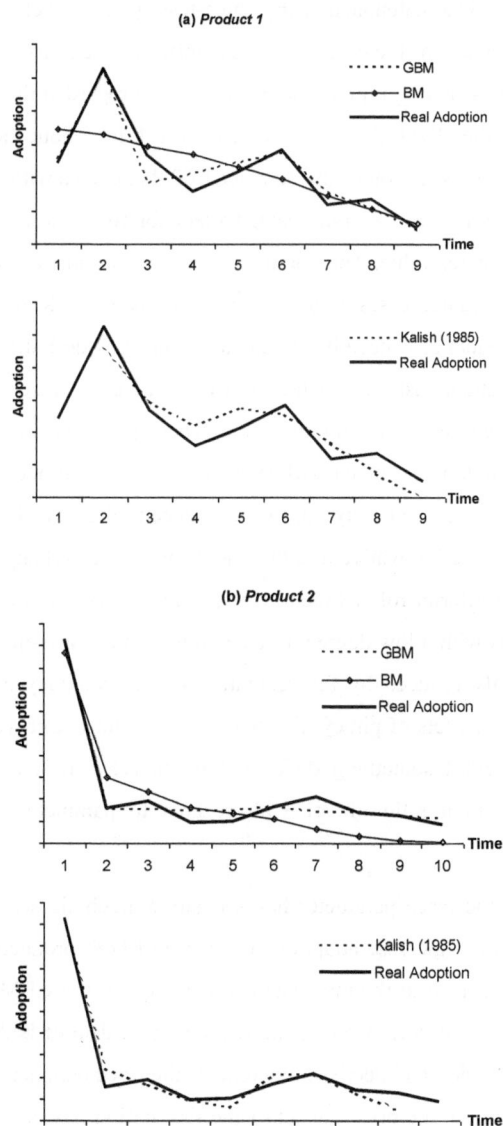

5.6.3   Statistical Significance and Plausibility

The significance of the estimators was calculated using the t-test with a level of significance of $\alpha = 0.05$. Not all parameters of the estimation are significant. Reasons for this is that both products did not yet pass a long time in the market, so the regarded time is relatively short. The estimation of the GBM along the product 1 did bring about the most significant estimators though. With exception of the parameter $\beta 2$ that comes with a high asymptotic standard error, all parameters are significant. Except for five cases, all parameters are plausible in their parameter value. The parameter q (adoption impact by personal communication) is negative in three cases, with both estimations of the Kalish model and with the BM relating to product 2. A negative parameter value is by definition not in accordance to the models theoretical basis. The values, however, are close to zero. It says that they bar the impact through personal communication. These parameters did not prove significant in the t-test. By looking at the real diffusion curves of both products, you can see that the peak sale is always reached early on. In accordance to the diffusion literature, it means that the consumers are innovative and that the information exchange by personal contact does only play an inferior role (Rogers, 1995). Such a curve is often observed in the context of innovations with a low degree of uncertainty to the consumer. The uncertainty might be significantly reduced by the reputation of the company as discussed before. Please note that due to acts of piracy, this hypothesis might not prove correct. The results of the GBM suggest something different though. Here, it is the parameter q that drives adoption with a ratio of 7:1 with regards to parameter p (innovation coefficient).

In both estimations of the GBM, the price parameter has a positive algebraic sign. According to this, a price increase would increase adoption. The price of both products did not change during our observation. Due to the mathematical equation of the GBM, the price only impacts adoption in the first period. Since the real adoption data of both products is relatively high in the first period and decreases thereafter, the price parameter is mapped wrongly. Within the estimation, the price impact of the first period is seen as the essential factor raising adoption, and therefore has a positive algebraic sign. The mathematical mapping of the price of the GBM thus leads to a wrong mapping and to the wrong conclusion of a positive coherence between price and adoption. All estimations of the market potential are plausible, since they are higher than the cumulated adoption within the time span. Nevertheless, the estimations do not exceed the current adoption by far. Since the trend of the real data (real adoption) is decreasing, the market potential is quiet small in both cases. Lilien et al. (1999) do not recommend using the NLS method of estimation with regards to market potential, since it calculates it too small most of the times, especially if there is only little data.

5.6.4   The Prognosis

A prognosis for the subsequent period is retrieved by substituting the estimated parameters into the equation of the model and by defining the parameter value of the decision variables. The price did not change in the subsequent period for both products; the advertisement spending was relatively high at this stage. The mean absolute percentage error (MAPE) is a statistical criterion suited to define the precision of the prognosis (Kamakura & Balasubramanian, 1987). Table 5.C show the MAPE.

Table 5.C: Quality of Prognosis

	GBM	BM	Kalish
Product 1: MAPE in $t = 10$	51,24%	58,88%	48,99%
Product 2: MAPE in $t = 11$	3,48%	98,26%	73,91%

The real adoption of product 1 in the period $t = 10$ was nearly twice as high as the real adoption of the preceding period. This shows how high the demands of reality on these models are. Even with a high level of advertisement spending, the level is not reached by the GBM or the Kalish model. The general trend of the preceding periods was a decrease. All three models have a percent deviation of around 50% with regards to adoption of product 1. Product 2 thereby only raises its real adoption from $t = 10$ to $t = 11$ by around 30%. Here, the GBM reaches a by far better prognosis. The prognosis deviates only by 3.48% from the real adoption. The comparison of the prognosis of the BM and the GBM shows that the incorporation of decision variables makes sense. The coefficient of determination of the BM is with 0.866 quiet high, the prognosis however is due to the lag of flexibility of the structure of the model with a MAPE of 98% poor. The model of

Kalish deviates by 73% and does not reflect the upwards trend of adoption. The GBM reflects the increased advertisement spending better than the Kalish model.

5.7 Discussion and Conclusion

Reasons for the strong deviation of the prognosis of the models can be seen in the framework of the estimation. The amount of data is quite small for calibration with only 9 (product 1) and 10 months (product 2). The estimation then is highly sensitive with regard to few alterations of the data (Bass, 1969). This data setting has been chosen because it shows one grave shortcoming of the quantitative diffusion research when seen with regards to applicability in business practice. A prognosis has the biggest utility before or shortly after rollout of an innovation. At this time, there is no or only little data available. Small amounts of data might lead to high deviations and non-significant parameters. High quantity of data minimises the deviation and shapes a more valid picture of the evolution of an innovation (Lilien et al., 1999). However, from a company's point of view, the utility of the insight is reduced by that time. Still, the incorporation of decision variables is justified by getting a better fit and prognosis. The model of Kalish cannot be compared to the BM, but shows that alternative ways to go about diffusion quantification with alternative theoretical founding contribute to a better understanding of the diffusion process. There is sufficient room for further research in the direction of monopoly markets becoming oligopolies, activities of competition with regards to branding and reputation of companies.

The market analysis with regard to factors influencing diffusion shows the high demand reality makes on these models. Incidents like piracy disrupt normal supply and demand patterns and a good reputation of companies might lead to diffusion shapes different to the widely acknowledged S-shaped curve. The application of these models to true data is scarce especially to data sets that deviate strongly from common assumptions. This analysis thereby suggests new challenges and interesting research spots. Future research might look deeper into shadow diffusion, counteractive measures such as the bundling of software products to hardware and how these measures influence diffusion. Goods are increasingly digitised and the underlying business logic and distribution are reconditioned. The models need to adapt changes in how products and services are distributed. The biggest challenge is the normative application of these models. A better prognostic capability is desirable, but the aim is not a prediction of the future. Sensitivity analysis and guidelines for optimal marketing strategies under varying market situations are more realistic and therefore desirable normative application areas for these models. Such normative specifications should be more widely considered in upcoming research. However, deficient estimation methods also complicate normative applications of these models (Lilien et al., 1999). The empirical test shows that there is a deficit in estimation of market potentials. This is consistent with a series of publications that claim an ex-ante specification of the market potential. However, the market analysis shows that for descriptive applications these models already contribute widely to uncover driving factors of diffusion. Further normative adjustments and consolidation of insights from the wide array of publications are eligible.

References

Bass, F. M. 1969. A New Product Growth for Model Consumer Durables. *Management Science*, 15: 215-227.

Bass, F. M. 1980. The Relationship between Diffusion Rates, Experience Curves, and Demand Elasticities for Consumer Durable Technological Innovation. *Journal of Business*, 53: 551-567.

Bass, F. M., Jain, D. P., & Krishnan, T. V. 2000. Modeling the Marketing-Mix Influence in New Product Diffusion". In V. Mahajan, E. Muller, & Y. Wind (Ed.), *New-Product Diffusion:* 99-123. Boston: Kluwer Academic Publishers.

Bass, F. M., Krishnan, T. V., & Jain, D. C. 1994. Why the Bass Model Fits without Decision Variables. *Marketing Science*, 13: 203-223.

Bottomley, P.A., & Fildes, R. 1998. The Role of Prices in Models of Innovation Diffusion. *Journal of Forecasting*, 17: 539-555.

Chow, G. C. 1967. Technological Change and the Demand of Computers. *American Economic Review*, 57: 1117-1130.

Dockner, E., & Jorgensen, S. 1988. Optimal Pricing Strategies for New Products in Dynamic Oligopolies. *Marketing Science*, 7: 315-334.

Gatignon, H., & Robertson, T. S. 1991. Innovative Decision Processes. In H. Kassarjian, & T. Robertson (Ed.), *Handbook of Consumer Behavior Theory and Research:* 316-348. Englewood Cliffs, NJ: Prentice-Hall.

Givon, M., Mahajan, V., & Muller, E. 1995. Software Piracy: Estimation of Lost Sales and the Impact on Software Diffusion. *Journal of Marketing*, 59: 29-37.

Hewing, M. 2012. A Theoretical and Empirical Comparison of Innovation Diffusion Models Applying Data from the Software Industry. *Journal of the Knowledge Economy*, 3(2): 125-141.

Horsky, D., & Simon, L. S. 1983. Advertising and the Diffusion of New Products. *Journal of Marketing Science*, 2: 1-17.

Jain, D., Mahajan, V., & Muller, E. 1991. Innovation Diffusion in the Presence of Supply Restrictions. *Marketing Science*, 10: 83-90.

Kahneman, D., & Tversky, A. 1979. Prospect Theory: An Analysis of Decision Under Risk. *Econometrica*, 47: 261-293.

Kalish, S. 1985. A New Product Adoption Model with Price, Advertising, and Uncertainty. *Management Science*, 31: 1569-1585.

Kalish, S., & Sen, S. K. 1986. Diffusion Models and the Marketing Mix for Single Products. In V. Mahajan, & Y. Wind (Ed.), *Innovation Diffusion Models of New Product Acceptance:* 87-115. Cambridge, MA: Ballinger Publishing Company.

Kamakura, W. A., & Balasubramanian, S. K. 1987. Long-Term Forecasting with Innovation Diffusion Models: The Impact of Replacement Purchases. *Journal of Forecasting*, 6: 1-19.

Kamakura, W. A., & Balasubramanian, S. K. 1988. Long-term view of the Diffusion of Durables. *International Journal of Research in Marketing:* 1-13.

Liftin, T. 2000. *Adoptionsfaktoren – Empirische Analyse am Beispiel eines innovativen Telekommunikationsdienstes.* Deutscher Universitäts-Verlag: Wiesbaden.

Lilien, G., Rangaswamy, A., & Van den Bulte, C. 1999. *Diffusion Models: Managerial Applications and Software.* Working Paper, University of Pennsylvania..

Mahajan, V., & Muller, E. 1979. Innovation Diffusion and New Product Growth Models in Marketing. *Journal of Marketing*, 43: 55-68.

Mahajan, V., Muller, E., & Bass, F. M. 1990. New Product Diffusion Models in Marketing: A Review and Directions for Research. *Journal of Marketing*, 54: 1-26.

Mahajan, V., Muller, E., & Bass, F. M. 1995. Diffusion of New Products: Empirical Generalizations and Managerial Uses. *Journal of Marketing Science*, 14: 79-88.

Mahajan, V., & Peterson, R. A. 1978. Innovation Diffusion in a Dynamic Potential Adopter Population. *Management Science*, 15: 1589-1597.

Putsis, W. P., & Srinivasan, V. 2000. Estimation Techniques for Macro Diffusion Models. In V. Mahajan, E. Muller, & Y. Wind (Ed.), *New-Product Diffusion:* 263-294. Boston: Kluwer Academic Publishers.

Raman, K., & Chatterjee, R. 1995. Optimal Monopolist Pricing Under Demand Uncertainty in Dynamic Markets. *Management Science*, 41: 144-162.

Rao, R. C., & Bass, F. M. 1985. Competition, Strategy, and Price Dynamics: A Theoretical and Empirical Investigation. *Journal of Marketing Research*, 22: 283-296.

Rogers, E. M. 1995. *Diffusion of Innovations.* 4th edition. New York: The Free Press.

Selnes, F. 1993. An Examination of the Effect of Product Performance on Brand Reputation, Satisfaction, and Loyalty. *European Journal of Marketing*, 27: 19-35.

Simon, H. 1982. ADPULS: An Advertising Model with Wearout and Pulsation. *Journal of Marketing Research*, 19: 352-363.

Simon, H. 1992. *Preismanagement.* 2nd edition. Wiesbaden: Gabler.

Srinivasan, V., & Mason, C. H. 1986. Nonlinear Least Squares Estimation of New Product Diffusion Models. *Marketing Science*, 5: 169-178.

Teng, J.-T., & Thompson, G. L. 1983. Oligopoly Models for Optimal Advertising. *Management Science*, 29: 1087-1101.